W9-DFR-692

The Misinformation Age

How False Beliefs Spread

Cailin O'Connor
James Owen Weatherall

Yale UNIVERSITY PRESS
New Haven & London

Published with assistance from the foundation established in memory of Calvin Chapin of the Class of 1788, Yale College.

Yale University Press books may be purchased in quantity for educational, business, or promotional use. For information, please e-mail sales.press@yale.edu (U.S. office) or sales@yaleup.co.uk (U.K. office).

Set in Janson Roman type by Integrated Publishing Solutions.
Printed in the United States of America.

Library of Congress Control Number: 2018940288
ISBN 978-0-300-23401-5 (hardcover : alk. paper)

A catalogue record for this book is available from the British Library.

This paper meets the requirements of ANSI/NISO Z39.48-1992 (Permanence of Paper).

10 9 8 7 6 5 4 3 2

To Eve and Vera

Contents

Note to Reader ix

INTRODUCTION
The Vegetable Lamb of Tartary 1

ONE
What Is Truth? 19

TWO
Polarization and Conformity 46

THREE
The Evangelization of Peoples 93

FOUR
The Social Network 147

Notes 187
Bibliography 215
Acknowledgments 251
Index 253

Note to Reader

Throughout the text, we use the plural pronouns "they" and "them" to refer to individual (i.e., singular) "agents" in mathematical and computational models of social learning. This usage follows the practice in the relevant academic literature. Although it may seem strange to some readers, the idea is to both (1) avoid assigning a gender to an abstract entity and (2) preserve the sense of "agency" of those entities, which would be lost if one were to use the singular "it."

The Misinformation Age

INTRODUCTION

The Vegetable Lamb of Tartary

In the middle part of the fourteenth century, a text purporting to be the travel memoirs of an English knight named Sir John Mandeville began to circulate among the learned of Europe.[1] Mandeville, the text claimed, had traveled through Asia Minor, northern Africa, and into India and had experienced many things unknown in Western Europe. Among these wonders was an Indian tree bearing gourdlike fruit, within which could be found tiny lambs, complete with flesh and blood. Mandeville claimed to have eaten this fruit and to have found it "wondirfulle."

He was not the only writer of the period to comment on these strange plants. An Italian friar named Odoric, who had also traveled through the East, wrote of a similar experience about thirty years before Mandeville.[2] He did not claim to have eaten the fruits but had learned of them on his travels from "persons worthy of credit."[3]

These reports of lamb-plants, which came to be known as the Vegetable Lamb of Tartary, captured the medieval imagination. The Vegetable Lamb was reported as fact by leading naturalists and botanical scholars, several of whom claimed to have studied it and to have seen its wool. These claims persisted into the seventeenth century, and books reported studies of plant-animal and animal-animal hybrids of which the Vegetable Lamb was merely one famous example. (Another example of a nonexistent beast that appeared in scientific texts was a horned hare that looked strikingly like a jackalope—the mythical creature of the American Southwest said to be fond of whiskey and able to imitate the voices of cowboys.)[4]

Even during its heyday, the Vegetable Lamb had its skeptics. Baron von Herberstein, a sixteenth-century ambassador to Russia from the Holy Roman Empire, reported that the Vegetable Lamb initially struck him as fabulous.[5] But as he looked into the matter, he gradually became a believer, particularly after so many "people of credence" described their direct experiences with the delicious flesh and downy snow-white wool of these lamb-pods. The various stories he heard "differed so little" that he came to believe "that there was more truthfulness in this matter than he had supposed."

Of course, there is no such plant and never was. Nor are there other such hybrids. But it took nearly four centuries after Mandeville's writings appeared for European botanists and biologists to recognize the Vegetable Lamb for a myth. In 1683, a Swedish naturalist named Engelbert Kaempfer, on direct order from King Charles XI, undertook a systematic search in Asia Minor and established conclusively that there simply are no Vegetable Lambs in the world.

How could it happen that, for centuries, European scholars could assert—with apparent certainty and seriousness—that lambs grew on trees? How could a belief with no supporting evidence, a belief that should have appeared—given all available experience concerning both plants and animals, and, indeed, regular exposure to lambs—simply absurd, nonetheless persist for centuries?

More important, are the mechanisms by which such beliefs were formed and spread, even among so-called experts, still present?

What are today's Vegetable Lambs, and how many of us believe in them?

On September 26, 2016, about six weeks before Donald Trump was elected president of the United States, a website calling itself ETF News (endingthefed.com) posted a story with the headline "Pope Francis Shocks World, Endorses Donald Trump for President, Releases Statement." The story included a statement allegedly from the Pope, asserting that he had decided to endorse Trump because the FBI had not pursued criminal charges against his opponent, Hillary Clinton.[6] (FBI director James B. Comey had announced on July 5, 2016, that the FBI had concluded an investigation of Clinton's use of a personal email server while secretary of state. Because the investigators found no evidence of intent to violate any law, they did not recommend prosecution.)[7]

"The FBI . . . has exposed itself as corrupted by political forces that have become far too powerful," said the Pope, according to ETF News. "Though I don't agree with Mr. Trump on some issues, I feel that voting against the powerful political forces that have corrupted the entire American federal government is the only option for a nation that desires a government that is truly for the people and by the people." The article was shared or liked on Facebook 960,000 times between when it was posted and the election. (The total number of users who *saw* the article, and even clicked on it, is perhaps ten times higher.)[8] It was the single most-shared election-related news item on Facebook in the three months leading up to the election.[9] By contrast, the most-shared article from a reputable news source during the same period was a *Washington Post* opinion piece titled "Trump's History of Corruption Is Mind-Boggling. So Why Is Clinton Supposedly the Corrupt One?" It was shared 849,000 times. (Of course, this piece, too, was nakedly partisan and hardly "news" in the standard sense—though the reporting in the article met much higher journalistic standards than the ETF News fabrication.) Also by way of contrast: Clinton lost eighteen US states by fewer than 250,000 votes—and she lost Wisconsin, Michigan, and Pennsylvania by fewer than 50,000 votes each. Had these three states flipped, Clinton would have won the election.[10]

Had the Pope endorsed Trump, it would have been major news, covered widely by many news organizations and surely deserving of attention. But it never happened. The entire story was fabricated.

The papal endorsement was the biggest "fake news" story of the election cycle, but it was hardly an outlier. An analysis by Craig Silverman at BuzzFeed News found that the top twenty fake news stories in the three months before the election were shared or liked a total of 8.7 million times on Facebook. Over the same period, the top twenty news stories from reputable sources got only 7.3 million Facebook shares or likes. In another study, economists Hunt Allcott at New York University and Matthew Gentzkow at Stanford compiled a database of 115 pro-Trump and 41 pro-Clinton fake news stories that, together, were shared 38 million times in the weeks before the election. These shares, they estimated, led to hundreds of millions of click-throughs. They also produced a list of fake news sites that, together, had 159 million visits between October 8 and November 8.[11]

ETF News was particularly successful: five of its stories made the top twenty list, and together those stories were responsible for 3.5 million shares or likes during this period.[12] ETF was arguably the most popular "news" source on the internet during the months before the election. Its stories included accusations that Clinton had sold weapons directly to the Islamic State, that US federal law disqualified her from holding office, that former FBI director James Comey had received millions of dollars from the Clinton Foundation, and that President Barack Obama had cut the pay of US military personnel.

In one noteworthy case, an ETF News story about the television host Megyn Kelly became a "trending topic" on Facebook and was actively promoted by the site. The article claimed that Kelly had been exposed as "a traitor" and fired from Fox News for supporting Hillary Clinton.[13] (A feud between Trump and Kelly had begun at a Republican primary debate on August 6, 2015, when Kelly asked Trump a pointed question about his past derogatory statements about women. Trump later made remarks that were widely interpreted as suggesting that Kelly's criticisms were due to her menstrual cycle.)

None of these stories was true. Many were not even original but were lifted directly from other fake news websites. The papal endorsement story, for instance, was originally posted on a site called WTOE 5 News, part of a network of fake news sites that drove traffic by claiming that celebrities were planning to move to various small towns in the United States.[14] (This site no longer exists.) Other articles were lifted, verbatim and without attribution, from such sites as supremepatriot.com and proudcons.com. Somehow, though, the versions posted on ETF News were shared far more on social media than the originals.

The persistence and spread of belief in the Vegetable Lamb was ultimately a harmless historical curiosity. But today's Vegetable Lambs have become a major political force not only in the US but also in the UK and in Europe.[15] Nearly a billion people live in the United States and the European Union (EU), and billions more are affected by the military, trade, and immigration policies of those nations. Whatever one thinks about the merits of Trump's election, or of the UK's exit from the EU ("Brexit"), it is profoundly troubling to think that these momentous political events were underwritten by falsehoods. And it raises a deep and unsettling question: Can democracy survive in an age of fake news?[16]

This is a book about belief. It is a book about truth and knowledge, science and evidence. But most of all, it is a book about *false* beliefs.[17] How do we form beliefs—especially false ones? How do they persist? Why do they spread? Why are false beliefs so intransigent, even in the face of overwhelming evidence to the contrary? And, perhaps most important, what can we do to change them?

This may sound like a truism, but it is worth saying: Our beliefs about the world matter. They matter to decisions we make every day. Do you eat sushi while pregnant? Well, that depends: Do you believe that the omega-3 fatty acids in fish will promote your baby's brain development? Do you believe that mercury in the fish will harm it? How likely is it that this particular restaurant harbors *Listeria*?

Beliefs also matter to decisions we make as a society—decisions concerning economic policy, public health, and the environment, among other topics. Do we restrict automobile emissions? This depends on our beliefs about how these emissions will affect public health and how restrictions will affect economic growth. Do we limit government debt? It depends on what we believe about whether that debt will affect our future well-being.

These sorts of beliefs are not idle, precisely because the decisions we make on their basis have real consequences. If you believe false things about the world, and you make decisions on the basis of those beliefs, then those decisions are unlikely to yield the outcomes you expect and desire. The world pushes back. If you do not believe that large ocean fish contain much mercury, or do not believe that mercury is harmful, then you may eat sushi while pregnant—and perhaps increase the chance of fetal mercury poisoning. If you believe that high government debt is a drain on the economy, even when interest rates are low, you will vote for policies that reduce debt, even at the cost of government services or stimulus spending.

That beliefs matter in this way means that holding false beliefs can hurt us. And that makes their staying power all the more startling.

To understand why false beliefs persist and spread, we need to understand where beliefs come from in the first place. Some come from our personal experience with the world. We have eaten tomatoes many times before, without apparent ill effects, so we believe they are not dangerous. (This, incidentally, is something Europeans did not discover for themselves for more than two hundred years; these "poison apples" were long considered dangerous to eat.)[18] We believe that regular exercise and consistent sleep will help our moods, that cold will cause chapped lips, and that gophers can be destructive pests in a garden. All these beliefs are based on direct prior experience with the world around us.

You might think that when we hold *false* beliefs—beliefs that are inconsistent with the available evidence, and which are even widely known to be inconsistent with that evidence—it is because of some failure to properly process the information we receive from the world.[19] Perhaps false beliefs are the result of cognitive biases or blind spots, quirks of human psychology that prevent us from drawing reliable inferences from our experience. Or else perhaps they come from a lack of experience or poor education. Or maybe people with false beliefs are simply too stupid to see the truth, even with the evidence right before their eyes. We often think of intelligence as a measure of just this: How effective is a person at drawing reliable inferences under various circumstances? False beliefs presumably indicate that a person is not drawing the right sorts of inferences.

Surely factors of this sort play a role in explaining how false beliefs form. But to focus on individual psychology, or intelligence, is to badly misdiagnose how false beliefs persist and spread. It leads us to the wrong remedies. Many of our beliefs—perhaps most of them—have a more complex origin: we form them on the basis of what other people tell us. We trust and learn from one another. Is there mercury in large fish after all? Most of us haven't the slightest idea how to test mercury levels. We must rely on information from others.

This is true of virtually every scientific belief we hold. If you ask most people in the United States whether the earth goes around the sun or vice versa, they will reply that it is the earth that moves. (At least, 74 percent will do so, according to a poll conducted by the National Science Foundation in 2012.)[20] They are (more or less) right. But they know this only because it is what someone told them. Even professional scientists tend to work on specific research topics addressing a miniscule subset of human knowledge and so are deeply dependent on others for the vast majority of their scientific beliefs. This means that even our society's most elite experts are susceptible to false belief based on the testimony of others. It was the scientists and thinkers of medieval Europe, remember, who were so knowledgeable about the Vegetable Lamb and its "wondirfulle flesh."

The ability to share information and influence one another's beliefs is part of what makes humans special. It allows for science and art—indeed, culture of any sort. But it leads to a conundrum. How do we know whether to trust what people tell us? If someone "of credence" tells you that indubitably the Vegetable Lamb is real—or that the Pope endorsed Donald Trump—do you believe that person? What if every smart, well-educated person you know believes in the Vegetable Lamb? What if your friends post articles about the Vegetable Lamb on social media and loudly joke about the ignorance of Vegetable Lamb Deniers? For most people, most of the time, this will be enough to convince them that, yes, there is probably more truth in the matter than they supposed.

When we open channels for social communication, we immediately face a trade-off. If we want to have as many true beliefs as possible, we should trust everything we hear. This way, every true belief passing through our social network also becomes part of our belief system. And if we want to minimize the number of false beliefs we have, we should not believe *anything*. That way we never fall for Vegetable Lambs. Of course, it would be best to believe only true things and never false ones, but for someone trying to adjudicate only on the basis of what he or she hears from others, it is hard to separate the true from the false. Most of us get our false beliefs from the same places we get our true ones, and if we want the good stuff, we risk getting the bad as well.[21]

Ultimately, what explains the persistence of the Vegetable Lamb in medieval texts has nothing to do with botany or the natural world. It was a purely social phenomenon. In thinking about Vegetable Lambs and fake news, we need to understand the social character of belief—and recognize that widespread falsehood is a necessary, but harmful, corollary to our most powerful tools for learning truths.

We live in an age of misinformation—an age of spin, marketing, and downright lies. Of course, lying is hardly new, but the deliberate propagation of false or misleading information has exploded in the past century, driven both by new technologies for disseminating information—radio, television, the internet—and by the increased sophistication of those who would mislead us.[22]

Much of this misinformation takes the form of propaganda. This is material produced, often by a government or political organization, to promote a particular viewpoint—or to challenge one.

Mass-media propaganda has long been a tool of governments to control their own citizens and to influence the political fortunes of their competitors, both domestically and abroad. American and European media consumers are regularly exposed to propaganda instruments produced by foreign adversaries such as the Russian-government-funded, English-language RT media organization and Sputnik News, which regularly reports on political events in the US and UK.[23] (They are also exposed to domestic government-sponsored media, such as the BBC in the UK.) And politically motivated media, ranging from news sources with a left or right "lean" to sources trading in conspiracy theories, rumors, and fake news, play a role much like that of state-run propaganda services.

Political propaganda, however, is just part of the problem. Often more dangerous—because we are less attuned to it—is *industrial* propaganda. This runs the gamut from advertising, which is explicitly intended to influence beliefs, to concerted misinformation campaigns designed to undermine reliable evidence.

A classic example of the latter is the campaign by tobacco companies during the second half of the twentieth century to disrupt and undermine research demonstrating the link between smoking and lung cancer. (We discuss this example in detail in Chapter 3.) Tobacco firms paid "experts" to create the impression that there was far more uncertainty and far less consensus than there actually was. This campaign successfully delayed, for a generation or more, regulation and public health initiatives to reduce smoking. As historians of science Naomi Oreskes and Erik Conway exhaustively document in their book *Merchants of Doubt*, the methods pioneered by cigarette makers have been emulated by the energy industry and allied scientists and politicians to create an impression of uncertainty concerning the severity and causes of climate change.[24]

All of these sources of deliberately partial, misleading, and inaccurate information—from political propaganda, to politically motivated media, to scientific research shaped by industrial interests—play an important role in the origins and spread of false beliefs. Just as it was Mandeville's fake memoirs that led to the widespread acceptance of the Vegetable Lamb, fake news and fake science remain crucial sources of false beliefs.

But the mere introduction of misinformation cannot explain its widespread acceptance. Unlike belief in the Vegetable Lamb, holding false beliefs about the health risks of smoking has serious consequences. And while a definitive case for a link between cancer and smoking requires large-scale epidemiological evidence and careful experiments, many, many people over the past century have watched loved ones—smokers—die premature, painful deaths. This is precisely the sort of direct experience that should bear on belief, unless other factors override it.

So how can propagandists override the weight of evidence from both direct experience and careful scientific inquiry to shape our beliefs?

We argue that propaganda often works by co-opting the same social factors that allowed the Vegetable Lamb—and the Pope's presidential endorsement—to persist and spread. In the case of cigarette smoking, those who wished to generate uncertainty could take advantage of the ways beliefs spread in a social network, and the human tendencies that regulate this process. The result was that millions of people suffered deaths that should have been preventable.

In this book we argue that social factors are essential to understanding the spread of beliefs, including—especially—false beliefs. We describe important mechanisms by which false beliefs spread and discuss why, perhaps counterintuitively, these very same mechanisms are often invaluable to us in our attempts to reach the truth. It is only through a proper understanding of these social effects that one can fully understand how false beliefs with significant, real-world consequences persist, even in the face of evidence of their falsehood. And during an era when fake news can dominate real news and influence elections and policy, this sort of understanding is a necessary step toward crafting a successful response.

In part, our argument draws on historical (and recent) examples of false beliefs that have spread through communities of people trying to learn about the world. Most of these examples, at least in the first chapters, come from science. We discuss how groups of scientists have come to hold false beliefs, and how those beliefs have persisted even as compelling evidence of their falsehood has appeared. We also discuss cases in which scientists have come to reject as false a belief they previously held. As we argue, scientists, just like the rest of us, are strongly influenced by their networks of social connections.[25]

But the fact that most of our examples come from science does not mean that our focus is limited to beliefs held by scientists. To the contrary: we wish to argue that the sorts of social considerations that we discuss here are crucial to understanding the persistence and spread of virtually *all* false beliefs. We focus on scientists because most scientists, most of the time, are doing their best to learn about the world, using the best methods available and paying careful attention to the available evidence. They are trained to gather and analyze evidence and they are generally well-informed about the issues they study. In other words, scientists are the closest we have to ideal inquirers. For these reasons, the fact that even communities of scientists can persist in false beliefs is striking—and if even scientists are influenced by social factors, surely the rest of us are as well.

There is another reason to focus on scientists. Ultimately, all of us, individually and collectively, need to act in the world, and the success of those actions depends on facts about how the world actually is. If we want to correctly anticipate what the consequences of our actions will be—which, surely, we do—then we want to carefully attend to the body of available evidence. When we do this, we often act like scientists: we try to learn what has happened in the past, to understand why things happen the way they do, and to predict what will happen in the future if we make various choices. We all have experiences, remember them, and change our beliefs in light of those experiences. Scientists, when doing science at least, merely try to be more systematic about this process. So we look to science as an extreme version of what all of us are doing much of the time as we try to make our way in the world.

Analyzing particular examples is only part of our strategy, however. The other part draws on methods from science itself: computer simulations and mathematical modeling. Over the past two decades, philosophers of science have taken models and ideas from fields such as economics and evolutionary biology and applied them to understand "epistemic communities"—that is, communities of people trying to gain knowledge about the world. Philosophers of science working with these models have tended to take communities of scientists to be their primary target, but as we argue later in the book, the models they have developed have much broader application: they can help us understand *any* community of people gathering evidence and sharing belief.

Why use mathematical models to understand something complex like human learning? Large-scale social effects can be very difficult to study using observational or experimental methods. This is because the processes at work are widely distributed through time and space and often involve hundreds or thousands of people. It is hard to intervene or to ask questions about how things would have turned out under slightly different circumstances in the way scientists usually do. This is one reason that simulations and modeling can help: by building simple computer programs that can mimic groups of people sharing ideas, it is possible to test hypotheses about what sorts of factors can matter to how such groups of people learn. This can then guide efforts to interpret what we see in the real world—and even suggest new ways of seeing the full complexities of human interaction.

Analysis of these models can change our basic conception of ourselves. There is a pervasive idea in Western culture that humans are essentially rational, deftly sorting fact from fiction, and, ultimately, arriving at timeless truths about the world. This line of thinking holds that humans follow the rules of logic, calculate probabilities accurately, and make decisions about the world that are perfectly informed by all available information. Conversely, failures to make effective and well-informed decisions are often chalked up to failures of human reasoning—resulting, say, from psychological tics or cognitive biases.[26] In this picture, whether we succeed or fail turns out to be a matter of whether individual humans are rational and intelligent. And so, if we want to achieve better outcomes—truer beliefs, better decisions—we need to focus on improving individual human reasoning.

Models of social learning help us see that this picture of human learning and rationality is dangerously distorted.[27] What we see in these models is that even perfectly rational—albeit simple—agents who learn from others in their social network can fail to form true beliefs about the world, even when more than adequate evidence is available. In other words, individually rational agents can form groups that are not rational at all.[28]

This sort of disconnect between individual and group-level rationality holds important morals for our understanding of human beliefs. Humans are animals that evolved to have the abilities they needed to thrive in their evolutionary environments. Most important, we are social animals. We have evolved to live in social groups, and to use these groups to share and create knowledge and understanding about the world. Our ability to successfully evaluate evidence and form true beliefs has as much to do with our social conditions as our individual psychology.

Taken together, these models, supplemented with evidence from other disciplines, provide a startling picture of our (in)ability to process information and make decisions. They help explain the Vegetable Lamb—but also explain how far more dangerous beliefs can take hold and spread.

In the 1964 political satire *Dr. Strangelove*, US Air Force general Jack Ripper unilaterally launches a nuclear strike on the USSR. His motive is to protect our "precious bodily fluids" from Soviet attack—an apparent reference to the belief, promulgated by the far-right John Birch Society during the 1950s and 1960s, that water fluoridation was a communist plot against America.[29] This sort of knee-jerk rejection, on the basis of pure speculation, of evidence-based public health interventions has long been a part of American public life. And yet, while the John Birch Society and their kin were certainly sincere in their opposition to water fluoridation (and many other government activities), they remained firmly on the fringes during the twentieth century.

Today, however, the situation appears to be different. Evidence-poor arguments about public-health issues such as global climate change, vaccination, and genetically modified foods are not only widely discussed and credited in mainstream political discussions, but in many cases they are actively supported by members of the current US administration, members of Congress, and some leading politicians in the UK, EU, and elsewhere. And as we have already noted, fake news and widespread false beliefs seem to have played significant roles in the 2016 US election, the UK Brexit vote, and other recent elections in Europe.

Of course, as we have already argued, to some degree the persistence of false beliefs is simply part of the human condition. The core structures of human knowledge and belief are such that social effects can lead to the spread of falsity, even in cases where the world pushes back.

And yet, there can be no doubt that the situation is changing. Over the past two decades, influential figures in American and British public life have adopted an ever-more-tenuous connection to the truth—and a complete disregard for evidence, expert knowledge, or logical coherence—with no political consequences. This leads to two urgent questions: What has changed? And how can we fix it?

One of our key arguments in this book is that we cannot understand changes in our political situation by focusing only on individuals. We also need to understand how our networks of social interaction have changed, and why those changes have affected our ability, as a group, to form reliable beliefs.

Since the early 1990s, our social structures have shifted dramatically away from community-level, face-to-face interactions and toward online interactions. Online social media such as Facebook and Twitter dramatically increase the amount of social information we receive and the rapidity with which we receive it, giving social effects an extra edge over other sources of knowledge. Social media also allows us to construct and prune our social networks, to surround ourselves with others who share our views and biases, and to refuse to interact with those who do not. This, in turn, filters the ways in which the world can push back, by limiting the facts to which we are exposed.[30]

Propagandistic tools are especially effective in this environment. Social commentators have long noted the emergence of online "echo chambers" in political discourse; models of social learning allow us to say why these changes matter so much.[31] People like to conform with those around them, and when we are surrounded by peers who hold identical beliefs, the forces of conformity become extremely strong. And as we argue, this tendency to conform can be weaponized—as evidenced by Russian interventions in the US and UK in 2016.

Likewise, the spread of ideas from scientists and other experts to the public and to politicians is deeply influenced by social factors—and for this reason, is readily manipulated. One of the most surprising conclusions from the models we study in this book is that it is not necessary for propagandists to produce fraudulent results to influence belief. Instead, by exerting influence on how legitimate, independent scientific results are shared with the public, the would-be propagandist can substantially affect the public's beliefs about scientific facts. This makes responding to propaganda particularly difficult. Merely sussing out industrial or political funding or influence in the production of science is not sufficient. We also need to be attuned to how science is publicized and shared.

Building on this understanding of the subtle and pernicious ways in which propaganda works, we go on to argue that the effects of propaganda can occur even in the absence of a propagandist. If journalists make efforts to be "fair" by presenting results from two sides of a scientific debate, they can bias what results the public sees in deeply misleading ways.[32]

We also show why some of the most obvious interventions to reduce the spread of propaganda and fake news, such as breaking down barriers that prevent the spread of reliable information between different communities, are unlikely to succeed. To learn from people whose views differ from ours, we need to be connected to them, but we also need to trust them enough to believe what they share. In a polarized political environment, that sort of trust is hard to come by.

There is, of course, no silver bullet for preventing the spread of lies and misinformation. We think the interventions most likely to succeed involve radical and unlikely changes, such as the development of new regulatory frameworks to penalize the intentional creation and distribution of fake news, similar to laws recently adopted in Germany to control hate speech on social media.[33] And perhaps even more is needed—up to and including a reengineering of our basic democratic institutions. Given the scant possibility (and the risks) of such changes, we point to ways in which journalists and social media firms—and each of us individually—can limit the spread of misinformation without needing to limit speech. We all want to avoid becoming unwitting propagandists for someone else's interests.

But as important as it is to identify particular interventions, it is equally important to understand the systems in which we are intervening. The models of social learning we present here give us a powerful framework for studying the consequences of possible interventions, and to predict, at least in a qualitative way, how specific changes to social networks might help or hurt. This sort of analysis goes far beyond hand-wringing about cultural trends to strike at the heart of how changing social dynamics can affect our beliefs.

It is crucial that we get a firmer grasp on this problem. Increasingly in the West—including both the United States and much of Europe—decisions are being made on the basis of lies and falsehoods. While it might seem that the solution is more information, this view is too limited. We have more information than ever before. Arguably, it is the abundance of information, shared in novel social contexts, that underlies the problems we face.

ONE

What Is Truth?

In May 1985, Joe Farman, Brian Gardiner, and Jonathan Shanklin, all scientists working for the British Antarctic Survey (BAS), reported a startling discovery.[1] According to their measurements, over the previous four years the quantity of ozone—a molecule made of three oxygen atoms—in the upper atmosphere over Antarctica had dropped precipitously when compared with measurements made by the same survey during the period 1957–1973. The change was most pronounced during the Antarctic spring, beginning around October. It seemed that as the continent warmed each year, a hole now formed in the layer of ozone covering the southern part of the globe.

This was a troubling discovery. The earth is constantly bombarded by radiation from space, including large amounts produced by the sun. Some of this radiation is energetic enough to destroy DNA and other cellular structures. But as was first recognized in the late nineteenth century, much of this high-energy radiation never reaches the surface of the planet.[2] Light—which is comparatively harmless—gets in, but anything much more dangerous gets absorbed in the upper atmosphere. By 1881, it was understood that ozone was responsible for this absorption; and by the late 1920s it was known that the entire earth is wrapped in a protective layer of atmospheric ozone, between ten and fifteen miles above ground, that shields us from being constantly irradiated by our own sun. This ozone layer is essential to life on earth. And now, it seemed, it was disappearing.

The BAS data were met with shock—but also skepticism. In addition to the land-based measurement devices that the BAS scientists used, the National Aeronautics and Space Administration (NASA) had a satellite in orbit monitoring global ozone levels. If this hole were real, it should have been glaringly obvious in the satellite data. But the satellite team had detected no significant changes.[3] The BAS data were flatly contradicted by this arguably more reliable data.

Besides, the BAS team's observations were theoretically impossible. Scientists had been studying ozone depletion mechanisms intensely for the previous fifteen years, ever since a Dutch atmospheric chemist named Paul Crutzen had shown that nitrous oxide, a component in many fertilizers, could reach the upper atmosphere and interact with ozone.[4] The most worrying mechanism for ozone depletion was discovered by Sherwood Rowland and Mario Molina in 1974, in work that would ultimately earn them, along with Crutzen, a Nobel Prize for Chemistry.

Rowland and Molina were chemists at the University of California, Irvine. Their big discovery was that ozone depletion could come from a quotidian source: a class of chemicals known as chlorofluorocarbons (CFCs) found in household products such as refrigerators, air conditioners, and virtually all aerosols, from spray paint to underarm deodorant.[5] First synthesized in the late nineteenth century as a fire suppressant, CFCs were a marvel of modern chemistry: highly stable, nontoxic, and broadly useful.

But some of the same properties that make CFCs so wondrous also make them dangerous—in unanticipated ways. Their stability means that they do not break down after they are released into the environment. Instead, they diffuse through the atmosphere and slowly creep above the ozone layer. Once there, they are exposed to high-energy radiation from the sun, which does finally break them down. And in breaking down they release other chemicals—most importantly, chlorine—that interact with ozone, removing it from the atmosphere.

Millions of tons of CFCs had been produced and freely released each year through much of the twentieth century. According to Molina and Rowland's estimates, much of it was still in the atmosphere, slowly creeping toward the ozone layer.

Molina and Rowland's work opened the research floodgates, leading to hundreds of studies examining ozone depletion from CFCs and other sources.[6] The results of these studies were clear: human activity could affect ozone levels in the upper atmosphere, with potentially disastrous consequences for life on Earth. In 1975, the US government formed a task force on Inadvertent Modification of the Stratosphere (IMOS) to study whether new regulations were needed. The task force announced later that year that, yes, heavily restricting the use of CFCs was imperative. The following year, the National Academy of Sciences produced two reports confirming the IMOS task force's basic findings. In 1977 the Food and Drug Administration announced a general ban on CFCs in a range of applications in the United States, to take effect beginning in 1979.

So by the time the BAS team released its findings, it was well-established that ozone was being depleted—and a regulatory framework was already in place, at least in the United States, to curb emissions of the chemicals thought to be responsible. International negotiations were under way to implement a worldwide ban. The response to Molina and Rowland's work had been fast and decisive. But for just this reason, it was widely believed that the problem of ozone depletion was well in hand. No one, including Molina and Rowland, believed that there was any *immediate* risk of holes opening in the ozone layer. There was no known chemical process by which ozone could deplete so rapidly.

This made the BAS findings all the more shocking. If they were correct, they showed that the possible risks of an abstract future were already upon us. But how could they be correct?

When the BAS study appeared, most people in the know thought it was wrong—and that the satellite data, which had not detected an ozone hole, were probably correct.[7] But it never hurts to double check, especially when two studies seem to disagree. So Richard Stolarski, a physicist working at NASA's Goddard Space Flight Center in Maryland, decided to revisit the satellite data for ozone levels over Antarctica.

Stolarski had done some of the earliest work on ozone depletion in the upper atmosphere. In the early 1970s, in collaboration with Ralph Cicerone, he had conducted a study for NASA in which he evaluated the likely effect of the space shuttle program on ozone levels. He and Cicerone were among the first people to focus on chlorine as a probable destroyer of ozone, a crucial step in later work on CFCs. Stolarski was thus in a particularly strong position to evaluate both the BAS and satellite data.

On careful reevaluation, he was surprised to find that the satellite *had* detected the ozone hole. But no one had noticed. The reason was that the measured levels were *so* low that the data-processing software had thrown them out as outliers—"bad" data points that were probably some sort of instrument glitch.

How could this happen? Any time scientists run an experiment involving complex electronic equipment that produces a lot of raw data, they need to design computer systems to help them process and analyze that data. This often involves "cleaning" the data to correct for known systematic errors and running statistical tests to extract the quantities the researchers are actually interested in. To borrow from the title of Nate Silver's recent book *The Signal and the Noise* (2012), they need to separate the signal from the noise. Designing software to do this requires a certain artfulness—and a lot of knowledge about the thing you are trying to measure. In this case, ozone concentrations had never been known to fall below a certain level, and there was no known process by which they could get that low. So the satellite team had designed its data-processing system to assume that any such data points were unreliable.

It turned out, then, that the satellite data were consistent with the BAS data after all—and that NASA had missed the ozone hole precisely because it was so far outside the range of what anyone believed possible. The ozone hole appeared to be real, and ozone depletion was not under control as everyone had thought. But there was still a puzzle: How could the theoretical expectations have been so far off?

NASA and the US National Oceanic and Atmospheric Administration rapidly organized two major expeditions to Antarctica, in 1986 and 1987, to measure the levels of possible culprits and to try to identify the processes that were producing such dramatic ozone loss.

The Antarctic expeditions revealed that the ozone hole resulted from a confluence of several factors—including some that no one had foreseen. One of the main contributors was the fact that the air above Antarctica is so cold that clouds there are composed of ice particles rather than water vapor. It turned out that these ice particles remove nitric acid from the air, which in turn allows the chlorine released by CFCs to persist longer, increasing ozone depletion.

Meanwhile, the continent's weather patterns have a distinctive character: powerful, frigid winds circle the South Pole, forming what is known as a polar vortex. This vortex traps the air over Antarctica so that ozone from other regions of the atmosphere cannot easily mix in, and the chlorine present there cannot easily disperse. This led to chlorine levels much higher than anyone predicted, with little chance for the ozone to be replenished from elsewhere.

The two Antarctic expeditions also resolved another issue: the ozone hole had indeed been caused by excess chlorine in the upper atmosphere—chlorine that could be traced directly back to CFCs. There could now be little doubt that human activity was capable of altering our environment at a massive scale, and so quickly that within decades we had substantially eroded our natural protection against harmful radiation from the sun, at least in one part of the world. The complex systems that enable life on our planet turned out to be perilously fragile.

In September 1987, before the second Antarctic expedition had even occurred, an international treaty known as the Montreal Protocol on Substances That Deplete the Ozone Level was established; it went into force on January 1, 1989, and was soon ratified by all members of the United Nations. The original Montreal Protocol called for production cuts of 50 percent by all nations producing CFCs.[8] Two years later, at a meeting in London, the protocol was revised to include a complete ban on CFCs and other chemicals known to release chlorine.

The nations of the world had acted definitively and with conviction. And they had done so on the basis of sound and exhaustive science. In the end, our scientific process did the best thing we could ask of it: it saved us all from space radiation.

The Gospel of John tells us that Jesus was brought before the Roman prefect of Judea, Pontius Pilate, after the Jewish leaders of

Jerusalem accused him of attempting to usurp Roman power and declare himself a king.[9] But when Pilate questioned him, Jesus demurred. He did not proclaim himself king but merely a witness to the truth.

Pilate's rejoinder: "What is truth?"

Pilate's response dismisses the very idea of "truth" as some ideal concerning the reliability and accuracy of our beliefs. His skepticism—not about any particular matters of fact but about the idea of "truth" itself—places him in a long tradition in Western thought, going back at least to the ancient Greek Skeptics, of questioning not only whether we can ever truly know anything about the world, but whether there are even "truths" out there to know.

But his response is equally part of a long history of those in power using this very philosophical tradition to undermine their critics. It was this tradition that George W. Bush's political advisor Karl Rove invoked when he said to *New York Times* reporter Ron Suskind, "We're an empire now [meaning the United States], and when we act, we create our own reality." Whether she knew it or not, Trump counselor Kellyanne Conway also invoked this tradition when she famously referred to the "alternative facts" that then–Press Secretary Sean Spicer had offered in a recent press conference.[10] And, lest anyone imagine "alternative facts" to be the exclusive domain of recent Republican administrations: recall that it was the Democrat Lyndon Johnson who, along with Secretary of Defense Robert McNamara, launched a massive escalation of the war in Vietnam in 1965 on the basis of outright lies about the status and prospects of the ongoing conflict.[11]

The idea of truth presents many old, difficult philosophical problems.[12] Can we uncover truths about the natural world? Are there reliable methods for doing so? Can we ever really know anything?

These might seem like questions for philosophers to worry about, hardly demanding of our attention. But in fact they are as central to everyday politics, business, and even life and death as any question one might ask. And as the imperial tradition running from Pilate to Trump suggests, those in power have long understood their importance.

As a scientific consensus emerged during the middle part of the 1970s that CFCs posed a serious risk to ozone levels, and US policy makers began to implement regulatory responses, the chemical industry pushed back. Led by DuPont, the massive American chemical manufacturer, industry representatives argued against doing anything. They sang a common refrain: it was too soon to act, because there was still too much uncertainty. DuPont placed ads in newspapers and magazines across the country, arguing that "there is no persuasive evidence" in favor of the Rowland-Molina claims that CFCs contributed to ozone depletion and asking, "Should an industry be prejudged and useful fluorocarbon products be destroyed before any answers are found?" Likewise, a 1975 op-ed piece in the industry magazine *Chemical Week* claimed, of the role of CFCs in ozone depletion, that "we're talking about a basically unknown effect on a little-understood phenomenon brought on by a debatable cause," and went on to conclude: "One fact is clear: We don't have the facts. We don't even know for sure whether there is a problem."[13]

This wait-and-see approach may seem judicious.[14] In 1975, although the evidence was sufficiently strong to persuade the IMOS task force that CFCs posed an imminent threat, many questions remained. Gathering more evidence was surely a good idea. Indeed, as we have seen, the 1980s would reveal that the scientific consensus of the 1970s had been deeply flawed: the danger was far *greater* than anyone had understood!

The problem was that the industry continued to call for more research, and for delayed action, irrespective of how much evidence came in. As late as March 1988, after the BAS findings showed the presence of the ozone hole, after Stolarski's review of the NASA satellite data confirmed the BAS data, and after the 1986 and 1987 Antarctic expeditions provided direct detections of by-products of CFC interactions with ozone, the CEO of DuPont wrote to the US Senate to declare that there was no need for drastic reductions.[15] By this point, it was hard to imagine what further evidence you could ask for. And yet the industry kept asking for more—for *certainty*.

DuPont's stance is reminiscent of an argument most famously associated with the eighteenth-century Scottish philosopher David Hume—though similar arguments were made by the ancient Greeks.[16] Suppose that, having observed some kind of regularity in the world, you would like to draw a general inference about it. For concreteness: Suppose you observe that the sun has risen every morning of your life. Can you infer that the sun always rises? Or, from the fact that you (growing up in the Northern Hemisphere, say) have only ever seen white swans, that every swan is white?

Hume's answer was an emphatic "no." No number of individual instances of a regularity can underwrite a general inference of that sort. This might sound like an absurd position, but the examples just offered illustrate the problem. There are black swans in Australia, after all, and there is no way to secure oneself against that possibility by checking more and more swans in Britain. The sun will eventually run out of hydrogen and expand to become a red giant, likely engulfing the earth. No matter how many days you have seen the sun rise, tomorrow could be the day it explodes.

This has become known as the "Problem of Induction."[17] Hume concluded that we cannot *know* anything about the world with certainty, because all inferences from experience fall prey to the Problem of Induction. The fact is that science can always be wrong.

Industry advocates urging a wait-and-see approach to CFCs into the late 1980s and beyond were right that the evidence linking CFCs to ozone depletion was not definitive. It still isn't. We cannot be absolutely certain about the existence of an ozone hole, about whether CFCs caused it, or even about whether ozone is essential for protecting human health. The reason we cannot be certain is that all of the evidence we have for this claim is ultimately inductive—and as Hume taught us, inductive evidence cannot produce certainty.

And it is not merely that we cannot be certain. Scientists have often been wrong in the past. The history of science is littered with crumpled-up theories that scientists once believed, on the basis of a great deal of evidence, but which they now reject. For nearly two thousand years, scientists believed bad air, or "miasma," emanating from rotting organic matter was the chief cause of disease—until the nineteenth century, when they came to believe that the diseases previously attributed to miasma are caused by microorganisms (i.e., germs). A thousand years of precision measurements and careful mathematical arguments had established, beyond a shadow of doubt, that the earth stands still and that the sun, planets, and stars all move around the stationary earth—until a series of scientists, from Copernicus to Newton, questioned and then overturned this theory. And then for centuries after that, Newton's theory of gravitation was accepted as the true explanation of the motions of the moon around the earth and the earth around the sun. But today even Newton's theory has been left behind for Einstein's theory of relativity.[18]

Philosophers of science, such as Larry Laudan and P. Kyle Stanford, have argued that these past failures of science should make us very cautious in accepting current scientific theories as true. Their argument is sometimes called the "pessimistic meta-induction": a careful look at the long history of scientific error should make us confident that current theories are also erroneous.[19]

Does this mean that industry critics of the scientific consensus on CFCs and ozone had a point? Scientists did not (*could* not) have enough evidence to be certain—and science has had such a dismal track record of discovering lasting truths that we can hardly take the scientists' word if they say they have gotten it right this time. Surely caution about accepting new scientific findings is always in order.

Not quite. Perhaps we can never be *certain* about anything, but that does not mean we cannot be more or less confident—or that we cannot gather evidence and use it to make informed decisions. We might, for instance, become very, very confident that CFCs are creating a hole in our ozone layer. With the right sorts of evidence we might become so confident that the line between this sort of evidentially grounded belief and absolute certainty is, for our purposes, meaningless.

Ultimately, we care about truth (at least scientific truth) inasmuch as true beliefs allow us to act successfully in the world. We care about knowledge because of the role that what we know—or at least, what we strongly believe to be true—plays in the choices we make, either individually or collectively. And recognizing this relationship between our beliefs and our choices is the key, not to solving the Problem of Induction, but to setting it aside.

When it comes to CFCs and the ozone layer, the worry that we can never gain complete certainty about matters of fact is irrelevant. What we want is enough confidence to avoid getting fried by radiation from space. When it comes to the question of what we should *do*, we need to set general skepticism aside and act on the basis of the evidence we have. We ignore demands for certainty from industry, and regulate. As Hume himself put it, "A wise man . . . proportions his belief to the evidence."[20]

These philosophical challenges to science can be reapplied to everyday life. There, too, we can never be certain. But the possibility does not paralyze us, nor should it. We do not wait on absolute certainty—and we cannot, as it is certainly not forthcoming. We have little choice but to act. And when we do, our actions are informed by what we happen to believe—which is why we should endeavor to have beliefs that are as well-supported as possible.

And on reflection, although scientists have come to reject many past theories, it remains true that those theories were often highly effective within the contexts that they had been developed and tested. The old earth-centered model of the solar system was supremely accurate for predicting the locations of stars and planets. We still use Newton's law of gravity to calculate satellite trajectories—and Newton's theory sufficed to get us to the moon. In other words, we make our beliefs as good as we can on the basis of the evidence we have, and, often enough, things work out.

Philosophers and statisticians over the past century and a half have developed ways of thinking about the relationship between belief, action, and evidence that captures this pragmatism.[21] The basic idea is that beliefs come in *degrees*, which measure, roughly, how likely we think something is to be true. And the evidence we gather can and should influence these degrees of belief. The character of that evidence can make us more or less confident. And when we make decisions based on our beliefs, we need to take those levels of confidence into account.

In fact, we can use a branch of probability theory to map out a precise relationship between what we observe and what we ought to believe. There is a formula, known as Bayes' rule, that allows you to calculate what your degree of belief, or *credence*, should be after learning of some evidence, taking into account what you believed before you saw the evidence and how likely the evidence was.[22] Bayes' rule is the unique, rational way to update your beliefs, in the sense that if someone does not use it, there will always be some series of bets that you could offer them, all of which they would want to take, but which they are certain to lose.

The formula itself does not matter for what follows. But the basic idea—that we can think of ourselves, and of scientists, as gathering evidence and updating our degrees of belief in light of it—will be very important later in this book.

Critics of the CFC ban had another argument against accepting the scientific consensus on ozone: the scientists who defended it were politically motivated. Sherwood Rowland was criticized particularly harshly. He and others like him were not "true, objective scientists," in the words of L. Craig Skaggs, DuPont's public affairs manager in the late 1970s, because they did not merely collect and report data: they also advocated for policy changes on the basis of that data.[23] Another industry executive put it more colorfully: criticism of CFCs was "orchestrated by the Ministry of Disinformation of the KGB." In other words, industry advocates argued, scientists whose work bears on political and industrial interests must be treated with extreme caution, since it is highly likely that their own political, economic, and even moral views will be reflected in the work that they do and the positions they adopt.

These arguments, much like industry's demand for certainty, reflect an important tradition of philosophy of science arguing that the work of scientists is shaped by sociological considerations, including their culture, their politics, and their values.[24] To understand this tradition, we need to turn to its roots, in the early 1960s.

In 1962, Thomas Kuhn, a physicist-turned-historian, published a book called *The Structure of Scientific Revolutions.*[25] Drawing on a litany of case studies in the history of physics, he described a pattern in scientific practice. Scientists would identify problems, apply well-known methods to solve them, run experiments to test their solutions, and gradually build their repertoire of bits of nature tamed by their work. Kuhn used the term "normal science" to describe this gradual progress.

Normal science may sound a lot like, well, *science*. But Kuhn's big insight was that it was only part of the story. Every so often, something else would happen: there would be a revolution. And when a revolution took place, Kuhn argued, the attendant changes in scientific practice were so dramatic that essentially none of the hard-won victories of the previous period of normal science would survive.

All normal science, Kuhn argued, occurs within some paradigm, with its own rules for identifying and solving problems and its own standards of evidence. As an example, today when we see a glass fall to the floor and shatter, we see an object pulled down by the force of gravity.[26] Before the paradigm of Newtonian gravitation, we did not see any such thing. We saw the glass as something made of earth, which therefore tended to move toward the earth, returning to its own level in a strict hierarchy of elements.

A scientific revolution is a change of paradigm: a radical discontinuity, not only in background theory, but in scientists' whole way of seeing the world. Changes of paradigm could change not only theory, but also what counts as evidence—and in some cases, Kuhn argued, even the results of experiments changed when paradigms changed.[27]

If taken to the extreme, Kuhn's ideas have some radical consequences.[28] Most work in philosophy of science before Kuhn viewed science as dispassionate and objective inquiry into the world.[29] But if Kuhn was right that paradigms structure scientists' worldviews and if all of our usual evidence gathering and analysis happens, by necessity, *within* a paradigm, then this picture was fatally flawed.[30] The "evidence" alone could not lead us to scientific theories. There was apparently another ingredient to science—one that ultimately had more to do with the *scientists* than with the world they were supposedly trying to understand.

Kuhn's work raised the possibility that to understand science, we had to recognize it as a human enterprise, with a complex history and rich sociological features that could affect the ideas scientists developed and defended. Scientists, from this perspective, were members of a society, and their behaviors were determined by that society's rites and rituals. More, their society was embedded in a larger cultural context. Understanding science would mean understanding this strange and novel culture, using the tools of fields such as sociology and anthropology.

Reconsidering science as embedded in a broader cultural and political context that could influence scientific thought led to some troubling realizations. Contemporary science had been produced and shaped by a largely male, white, and Western European culture that had committed atrocities around the world. And scientific ideas, it turned out, were implicated. For instance, historian Ruth Schwartz Cowan and sociologist Donald McKenzie outline how the whole field of statistics emerged when Karl Pearson and Francis Galton (Charles Darwin's cousin) attempted to quantify various markers of racial superiority.[31] (In fact, Galton coined the term "eugenics.") The French philosopher and historian Michel Foucault argued that modern psychiatry was an instrument of subjugation, a way of segregating "problematic" members of society from the rest of the population.[32] The modern clinic, he argued, was descended from the medieval leper colony and played a similar role in society.

Science was also implicated in colonialism, which had often been justified by "scientific" arguments about racial superiority and by the assumption that non-Western cultures could not have reliable knowledge about their own environmental and economic well-being.[33] In her 1986 book *The Science Question in Feminism*, Sandra

Harding, a prominent American feminist theorist, pointed to the proliferation of rape metaphors—with the scientist forcing an unwilling Mother Nature to submit—in the writings of early scientists, including the British empiricist Francis Bacon, who strongly influenced Newton (and Hume). And so on.

The 1970s and 1980s were a golden age for this tradition of work on science, politics, and culture, which came to be known as "science studies." And it was in this intellectual context that industry advocates raised the concern that scientists researching the ozone hole were themselves political agents, influenced by their background views about environmentalism, government regulation, and the value of industry. And for this reason, the criticism had some initial plausibility: after all, many scholars had argued that precisely this sort of cultural context *did* matter to science; if, in this particular case, a group of scientists seemed to be endorsing political views, might not those views have influenced their scientific work, in just the way that Galton's racism influenced the problems in statistics that he worked on?

It is true that, like all of us, scientists cannot isolate themselves from their cultural contexts. These contexts can surely lead to biases and blind spots. Sometimes the conclusions scientists draw in the grips of their own biases have been socially unacceptable, morally bankrupt, or just wrong, and yet were widely accepted nonetheless.[34] So it is true that we need to be aware of and sophisticated about how social and political factors may influence science—and it is for this reason that the sorts of cultural critiques of science emerging from science studies can and have been deeply valuable.

But the mere observation that a scientist or group of scientists holds certain cultural or political views does not undermine the evidence and arguments they produce to support those views. This is true even when those views have influenced the scientific problems the scientists work on or the methods they adopt. After all, despite its objectionable origins, the field of statistics is not inherently bankrupt because it was developed with particular goals in mind. The insights of Galton and Pearson have been developed into a large and invaluable set of tools for analyzing and interpreting data—even if the misuse of statistics remains commonplace.[35] Likewise, the hundreds of scientific articles written during the 1970s and 1980s providing careful evidence of the role of CFCs in ozone depletion, and the conclusion that CFCs posed an imminent threat to human health, are not washed away if some or even all of the authors of those articles happen to think that preserving the environment is an admirable goal in itself.

More, it is very important to distinguish between two ways in which politics might affect science. One consists in the sorts of subtle influences we have been considering here, wherein background cultural views affect the assumptions scientists make and the problems they consider. These sorts of influences can have negative effects, but they can also be identified and addressed through careful analysis, criticism, and debate. But there is another way in which politics and science can mix—one that has a strikingly different, and far more nefarious, character.

On Pentecost Sunday in 1783, a sixteen-mile-long volcanic fissure opened on the side of Laki, a mountain in southern Iceland.[36] Over the next eight months, tens of billions of tons of lava flowed out, engulfing twenty towns and villages. Lava fountains spouted molten rock nearly a mile into the air. Jón Steingrímsson, a local Lutheran priest who recorded the eruption, wrote that the "flood of fire flowed with the speed of a great river swollen with meltwater on a spring day."[37]

Along with the fast and destructive lava flows, the Laki eruption released enormous amounts of gas—including more than one hundred million tons of sulfur dioxide and nearly ten million tons of fluorine gas.[38] The gases reacted with atmospheric water vapor to form sulfuric and hydrofluoric acids—which quickly fell back to earth as rain. The rain was so acidic that it melted the skin and bones of local livestock, killing half of the island's horses and cows and three-quarters of its sheep. Barley and rye withered on their stalks. A quarter of Iceland's human population died in the resulting famine.

The expression "acid rain" was first coined in 1859 by Robert Angus, a British pharmacist working in Manchester.[39] Angus, who was studying sources of air pollution, found that the rain near industrial sites tended to be more acidic than that near the coasts, where there was less pollution. He attributed this effect to an early industrial technique known as the Leblanc soda process, which was known to release hydrochloric acid. Angus's research showed that rather than merely dissipate in the atmosphere, that acid tended to fall back to earth as acid rain.

As the Laki eruption demonstrates, the basic phenomenon of acid rain had already been rather dramatically observed. Angus's work showed that acid rain could also be a by-product of human activity. The British Parliament reacted quickly, implementing the Alkali Act in 1864, which required Leblanc process plants to prevent the acid from being released into the atmosphere. It was a remarkable early act of environmental regulation. And while acid rain did not go away after the Alkali Act passed, it did fade from public consciousness over the following century, perhaps because it seemed like a local problem for industrial regions to worry about.[40]

This all changed in 1974—the same year that Rowland and Molina discovered that CFCs deplete ozone. That year, Gene Likens, a professor at Cornell, and F. Herbert Bormann, a professor at Yale, published a research article in the scientific journal *Science* in which they defended a startling conclusion: the rain and snow falling on virtually the entire northeastern United States had become acidic—

much more acidic than elsewhere in the country, and more acidic than in the same region twenty years previously. Likens and Bormann reached this conclusion after carefully analyzing eleven years of data collected at the Hubbard Brook Experimental Forest in north-central New Hampshire and then comparing it with snow and rainwater samples from around the region.[41]

Likens and Bormann's conclusions were startling because they showed that even remote areas, far from industrial centers, were deeply affected by human activity and pollution. It seemed that sulfur and other chemicals released by power plants in the Ohio Valley were drifting over the entire Northeast before falling to the ground as acid rain. These findings were consonant with similar results in southern Sweden and in Norway, where acid rain was likewise observed far from industrial settings. (In this case, it seemed the pollution was coming from England and Germany.) Acid rain due to sources in the United States was soon detected in Canada. Long-distance pollution and acid rain were rapidly becoming not just an environmental problem, but a matter of international relations.

By the early 1980s, the science on acid rain was, in the words of one Environmental Protection Agency (EPA) spokesperson, "unimpeachable." This opinion was shared by virtually every major scientific organization tasked with reviewing the by-then enormous literature on the topic. In 1981, the National Academy of Sciences, the most prestigious group of scientists in the United States, found "clear evidence of serious hazard to human health" as well as "overwhelming" evidence that acid rain was caused by power-plant emissions. A 1982 report by the EPA concurred,[42] as did another National Academy report and a review by the Royal Society of Canada in 1983.[43] And so on.

This broad scientific consensus should have led to new limits on power-plant emissions. And in Europe, it did: in 1979, the United Nations Economic Commission for Europe passed a new convention to limit transnational pollution. That July, the United States and Canada began negotiating a similar agreement, leading to a Memorandum of Intent to address transnational air pollution.

But while these initiatives were started under the Carter Administration, the job of developing a full regulatory framework fell to Ronald Reagan, who became president of the United States on January 20, 1981. And despite the large body of evidence concerning the causes and harms of acid rain, the Reagan Administration did everything it could to prevent action—up to and including tampering with the scientific record.[44]

In 1982, George Keyworth, the White House science advisor, commissioned yet another report on acid rain. The purpose of the report was ostensibly to assess previous studies conducted under the Carter Administration.[45] The Reagan panel was called the Acid Rain Peer Review Panel and was headed by William Nierenberg, a distinguished physicist who had been director of the Scripps Institute of Oceanography, in San Diego.

Nierenberg's scientific credentials were beyond reproach—but he was also overtly partisan. Two years after taking on the Acid Rain Peer Review Panel, Nierenberg joined two other distinguished physicists—Frederick Seitz, the former president of both Rockefeller University and the National Academy of Sciences, and Robert Jastrow, founder of NASA's Goddard Institute for Space Studies at Columbia University—to start up a conservative think tank known as the George C. Marshall Institute.

Political alliances surely played a role in Nierenberg's appointment as head of the Reagan panel, but despite his political leanings, he took the role seriously. He hand-picked a team of distinguished experts on atmospheric science to join the panel—including Sherwood Rowland and Gene Likens. Six of the nine panel members were members of the National Academies of Sciences or of Engineering—and all but one of them was an expert on the relevant science. The final member was a physicist named Fred Singer.

According to Oreskes and Conway in their book *Merchants of Doubt*, when Nierenberg submitted a draft list of potential panel members to the White House, Singer's name was not on it. But when the White House sent the list back, it had crossed off several candidates—and added Singer. At the time, he held a tenured faculty position at the University of Virginia, but his primary affiliation seems to have been with the Heritage Foundation, a conservative think tank.[46] We could find no record of any research articles authored by Singer addressing acid rain before he was placed on the panel.[47]

By March 1983, the panel had produced a complete draft of its report, which was circulated among the members. Its findings concurred with those of every other serious group that had addressed the subject: acid rain was a real threat, it was caused by human activity, and drastic reductions in power-plant emissions were necessary to stop further damage. The draft also included a chapter written by Singer—in which he came to startlingly different conclusions. He argued that every course of action bore unacceptable costs, except doing nothing. The other members rejected this finding and refused to sign off on any version of the report that included Singer's chapter. The disagreement led to months of delays—during which time it appeared that Singer was trying to undermine the panel's main findings in other ways. In September 1983, for instance, the vice chair of the panel presented its consensus to Congress; Singer protested the testimony in a letter to Congress arguing that there was insufficient evidence to support the vice chair's claims.

The panel finally approved a version of the report in March 1984, a year after it was first produced. The final draft included Singer's chapter as a signed appendix without an endorsement from the panel. Even then it was not released but instead sent to the White House for further review; Keyworth's office then proposed further revisions to the Executive Summary portion of the document, which Nierenberg made without consulting with the rest of the panel. According to Kenneth Rahn, an atmospheric scientist at the University of Rhode Island and a panel member, these changes weakened "the panel's message that the federal government should take action now."[48]

The final report did not accurately reflect the panel's recommendations—and it was released, with panel members' names, in a form that those members had not reviewed or approved. This was a significant and controversial breach of protocol. *Science* magazine published an article charging that the White House had manipulated the report. Members of Congress released statements accusing the White House of suppressing the study, and several major newspapers covered the story.[49]

But the damage had already been done. In addition to watering down the report, Singer and Keyworth, with aid from Nierenberg, had delayed its release by more than a year. In the meantime, several bills had been introduced in Congress that would have addressed acidifying pollutants, but none of them had been taken up because the science of acid rain was, allegedly, still uncertain, pending the White House report.[50]

In the end, no legislation on acid rain was seriously considered for five more years, until after Reagan had left the White House.

When Sherwood Rowland won the Nobel Prize in 1995 for his work on CFCs and ozone, Fred Singer wrote an op-ed piece criticizing the Nobel committee: "In awarding the 1995 Nobel Prize in Chemistry to the originators of the stratospheric ozone depletion hypothesis, the Swedish Academy has chosen to make a political statement."[51] This critique echoed the earlier industry arguments that, by advocating for new regulations on the basis of his work on CFCs, Rowland had become a political agent rather than a "true scientist."

The irony here is deep. Even if we grant—as we surely must—that Rowland's work was influenced in various ways by his political and cultural contexts, there is a huge difference between speaking out about the socially relevant consequences of one's own scientific research, as Rowland did following his 1974 paper, and working in direct consultation with the White House to undermine a scientific panel's findings. The argument here is not *tu quoque:* it is not that Singer, too, was influenced by political considerations, just as Rowland surely was. Nor is it that generally right-leaning interference, of the sort exemplified by the Acid Rain Peer Review Panel, is at least as bad as potential left-leaning interference. Rather, the argument is that there is a kind of political interference in science that is apparent in the case of the Acid Rain Peer Review Panel, and which seems importantly different in kind from anything that advocates of CFC regulation were ever accused of: explicit and intentional manipulation of scientific reports.

There is a second irony here, as well, regarding social and historical studies of science. By the early 1990s there was a broad perception among many scientists, and also some philosophers, politicians, and journalists, that academics in the humanities were agitating to undermine science. These scientists began to push back. The result was a confrontation, ostensibly over the legitimacy of scientific knowledge, that came to be known as the "science wars."[52]

Perhaps the first volley in the science wars came in the form of a 1994 book called *Higher Superstition: The Academic Left and Its Quarrels with Science.* Written by biologist Paul Gross and mathematician Norman Levitt, *Higher Superstition* argued that the sociologists and philosophers who purported to analyze science were generally incompetent to evaluate the work they were responding to. The authors argued that many of the arguments in the science studies literature were not just ill-informed but downright incoherent.

One of the major themes of the science wars was an accusation that humanists writing on science were pseudointellectual poseurs.[53] But we want to focus on a second theme, which is suggested by the term "academic left" in the subtitle of Gross and Levitt's book. Their arguments suggested a deep intellectual connection between, on one hand, a certain kind of left-wing politics that emphasized diversity and multiculturalism, and on the other hand, a broadly antiscientific stance.[54] As the Harvard biologist Richard Lewontin put it in a letter to the *New York Review of Books* in 1998, Levitt and Gross appeared to hold the "curious belief . . . that any claim of social and ideological influence on the process and content of science is a form of Marxist madness."[55] This political message was amplified by prominent conservative commentators, so that the science wars soon came to look like a skirmish in a much larger culture war. Science was just one facet of a Western culture under siege by Marxists and, worse, in need of defense along with Christianity, figurative painting, and the collected works of Aristotle.

One of the ideologies that Gross and Levitt associated with the academic left—and criticized harshly—was what they called "radical environmentalism," personified by Jeremy Rifkin, an American author and activist who was famous for drawing attention to ozone depletion, acid rain, and global warming in the 1970s and 1980s.[56] Radical environmentalists, Gross and Levitt argued, were scaremongering pseudoscientists using the threat of environmental catastrophe to promote a (mostly leftist) political ideology.

And yet, in Rowland's case, it was the distinguished scientist and future Nobel laureate who, in speaking out about his own highly influential and widely cited research, was tarred as an activist and ideologue. Opposed to Rowland was not an apolitical and dispassionate scientific establishment, frowning on the activism of one of its members; it was a multibillion dollar industry with an army of lobbyists fighting regulation that had a direct impact on that industry's bottom line.

At the peak of the science wars, the claim that science had a political or cultural component was often presented as a radical assault on scientific authority. Reflection on the acid rain example suggests a very different moral. Yes, science is subject to political influence, both subtle and overt. But in a world where political appointees can rewrite scientific documents, without the knowledge and consent of the authors of those documents, fear that the public will stop trusting scientists because some historians or sociologists have pointed to problematic episodes in history seems wildly off the mark.

To the contrary, it was two historians, Oreskes and Conway, who carefully documented the Reagan Administration's interventions in the case of acid rain. Scholars working in disciplines caricatured as "antiscience" during the science wars have very often provided the critical analysis needed to set the record straight and improve both the methods and public understanding of science.[57]

And it was a Nobel Prize–winning "radical environmentalist" who led the charge against CFCs.

The picture of "truth" and "falsity" that we have sketched in this chapter is one according to which our beliefs play a particular role in guiding action.[58] We seek to hold beliefs that are "true" in the sense of serving as guides for making successful choices in the future; we generally expect such beliefs to conform with and be supported by the available evidence. We want to know that a layer of ozone in the stratosphere protects us from solar radiation, that ozone can be depleted by CFCs, and that that depletion has occurred so quickly in some places that a hole has opened in the ozone layer so that we are able to adopt regulatory frameworks to slow or reverse ozone depletion. Similarly for acid rain. (When we say, in what follows, that a belief is "true," this is all we mean; likewise, a "false" belief is one that does not bear this relationship to evidence and successful action.)

This picture is important to understanding how to think of scientific authority in the context of political debates. The real trouble is that most of us are not in a position to independently evaluate, much less collect and analyze, the full set of data that a given community of scientists can bring to bear on any particular problem. For that matter, most individual scientists are not in this position either! This means that if scientists claim they are gathering evidence and that evidence is convincing, we have little choice but to take their word for it. And whether we accept what scientists tell us depends on the degree to which we trust scientists to accurately gather and report their evidence, and to responsibly update their beliefs in light of it.

Ultimately, the reason to rely on scientific knowledge when we make decisions is not that scientists form a priesthood, uttering eternal truths from the mountaintop of rationality. Rather, it is that scientists are usually in the best position to systematically gather and evaluate whatever evidence is available. The views of scientists on issues of public interest—from questions concerning the environment, to the safety and efficacy of drugs and other pharmaceuticals, to the risks associated with new technology—have a special status not because of the authority of the people who hold them, but because the views themselves are informed by the best evidence we have.

We can see this clearly in the cases of both the ozone hole and acid rain. The reason to believe that there was a hole in the ozone layer was not because scientists *said* there was such a hole; it was because multiple devices, designed and carefully calibrated to measure ozone levels in the upper atmosphere, located in different places and making their measurements in different ways, detected substantially reduced ozone levels. And the reason to believe that CFCs had produced that ozone hole was because a carefully tested theory was available according to which CFCs could deplete ozone—and the predicted by-products of that process were detected in the atmosphere over Antarctica. Likewise for sulfur produced by coal-fired power plants and acid rain in distant regions.

All of this remains true even if we accept that science is deeply entwined with culture and politics—and that scientists have often gotten things wrong in the past. Seeking out the ways in which current science is flawed is a natural part of science itself. If we take seriously the possibility that we are wrong—a stance that is central to the whole scientific enterprise—then we should be eager to give weight to criticism, particularly of a sort that scientists themselves may miss. The upshot of this criticism will be better science: more-convincing arguments, fewer distortions, and better guidance to action. It is the hard work of serious sociologists, anthropologists, historians, and philosophers of science—and scientists themselves—that has helped us better understand how science works, and why it should play a central role in our decision making.

From this perspective, the real threat to science is not from the ways in which it is influenced by its cultural context, nor the philosophical and social critiques that draw those influences out. Rather, the real threat is from those people who would manipulate scientific knowledge for their own interests or obscure it from the policy makers who might act on it. We are good at dismissing philosophical worries and acting when necessary. We are much less good at protecting ourselves from those who peddle misinformation.

TWO

Polarization and Conformity

Elemental mercury is the only metal that is liquid at room temperature. Aristotle called it "quicksilver," a term that captures its strange beauty. But this particular beauty is also deadly. Exposure to mercury can lead to a host of symptoms: sensations of bugs crawling under the skin, extreme muscle weakness, hair loss, paranoia, mental instability, and, for high exposure levels, death.[1]

The history of mercury use is riddled with such poisonings. Qin Shi Huang, the first emperor of a unified China, is reported to have died in 210 BCE after taking mercury pills that ironically were intended to make him immortal.[2] Isaac Newton sank into paranoia and insanity at the end of his brilliant life—likely a result of his experiments with mercury. (Posthumous hair samples revealed highly elevated levels of it.)[3]

By the end of the twentieth century, the dangers of mercury were well established, and its use was heavily regulated in the United States, Europe, Japan, China, and elsewhere.[4] Mercury poisoning should have been under control.

And yet beginning around 2000, an American physician named Jane Hightower began to notice a distinctive cluster of symptoms in her patients: hair loss, nausea, weakness, brain fog. These are all associated with mercury poisoning—but these patients did not have lifestyles that should have brought them into contact with heavy metals, and so the diagnosis did not occur to her. Until, that is, a colleague heard a story on public radio about a town where locals suffered hair loss and other ailments of mercury poisoning after eating contaminated fish.[5] On a hunch, this colleague ordered a mercury test for one of Hightower's patients.

Sure enough, the patient's mercury levels were elevated.

The patient also ate a lot of fish. Armed with a new hypothesis, that the strange symptoms were linked to mercury and perhaps to fish, Hightower went back to her other mystery patients with a new question. How often did they eat fish? As it turned out, those patients tended to be wealthy and health-conscious and chose to eat fish *very* often—including many fish high on the food chain, such as shark, swordfish, and tuna.

Over the next few years, Hightower systematically recorded her observations and shared her suspicions with colleagues, including some EPA officials who worked on mercury contamination in seafood. Some of the doctors she spoke with began to look for evidence of mercury poisoning in their own patients. Obstetricians in her hospital warned pregnant women off certain fish, since fetal brains are particularly susceptible to the effects of mercury.[6] Some doctor friends quit eating predatory fish. The hospital cafeteria stopped serving canned tuna.

A local news station ran a story on Hightower's suspicions.[7] Then *20/20*, a national television news program, ran a segment on mercury poisoning and fish.[8] Television crews performed tests of the mercury levels in fish at local supermarkets and discovered that some of them, especially shark and swordfish, were well above levels deemed safe by the US Food and Drug Administration (FDA). Their coverage of Hightower's claims reached a wide audience—and soon more doctors were monitoring their patients for fish-related mercury poisoning, gradually accumulating a larger and larger body of evidence supporting Hightower's hypothesis.

We often associate scientific discovery with lone geniuses—mercurial madman Isaac Newton, Charles Darwin, Albert Einstein—who, in a moment of revelation, conceive of some new theory fully formed. But real discoveries are far more complicated and almost invariably involve many people.[9] Most scientific advances result from the slow accumulation of knowledge in a community. Guesses and observations come from many directions. These insights gradually spread and accumulate, leading to yet more hypotheses and new ideas for how to gather evidence. Only after a long and collaborative process can we say that scientists have achieved a new discovery. Crucial to this process is the network of human interaction linking scientists to one another.

Although Jane Hightower led the effort to link mercury poisoning with overconsumption of contaminated fish, she did not act alone. It was a colleague who first connected hair loss in Hightower's patient with mercury poisoning. It was a contact at the EPA who, upon hearing about her work, shared recent government studies on mercury in fish. Other doctors informed her of patients with similar symptoms, improving her understanding of the syndrome. Hightower's thinking was informed at every step by evidence from outside her own experience.

Conversely, Hightower's insights helped others make even more progress. As soon as she started to gather evidence, her work began influencing the beliefs and behaviors of those around her—obstetricians, other clinicians, medical associations—who went on to find more evidence and further links. Ultimately, the discovery of a new link between mercury poisoning and seafood consumption occurred when a community, or *network*, of scientists and doctors, all sharing ideas and evidence, adopted a new consensus.

In this way, those responsible for scientific discovery are bolstered by those around them. Bolstered—but also, sometimes, stymied. Hightower's evidence did not convince everyone she shared it with. To many colleagues, she seemed like an activist with some kind of environmentalist axe to grind, or perhaps just a quack. In fact, there seemed to be good reasons to think that the symptoms Hightower observed could not be from mercury.

In the early 2000s, it was already widely known that some fish contained mercury. Coal-fired power plants emitted a form of inorganic mercury into the air, where it would gradually fall back to earth, mix into ocean water, and be ingested by microbes, which converted it to highly toxic methylmercury. These microbes would then be consumed by small fish, which would be consumed by larger fish, and so on up the food chain. Methylmercury tends to accumulate in animal tissue, so large fish were building up high levels of the toxin. This was why the FDA already had guidelines regulating the level of methylmercury in fish sold commercially—levels that, it turned out, were exceeded by some supermarket supplies.

So the idea that fish contained toxic mercury was not controversial. But precisely because the whole process seemed well-understood, regulators, including the FDA, thought they knew what the dangers were. When presented with Hightower's work, the FDA responded that no one was actually eating enough fish to be poisoned. Many of her colleagues seemed to agree.

Still, Hightower pushed forward with a year-long survey documenting the fish intake, symptoms, and blood mercury levels of a group of patients. She published these results and shared them with a contact at the EPA, who invited her to present her work at a meeting of mercury experts. At the suggestion of another colleague she wrote a resolution about the dangers of methylmercury and how to tackle them, which was passed by the California Medical Association and San Francisco Medical Society.

With time and ever more evidence, she gradually convinced more and more of her colleagues. Today, government agencies around the world are more savvy about the risks of methylmercury poisoning from fish and have issued guidelines to better control exposure.

On February 28, 1953, around lunchtime, the English biologist Francis Crick called for the attention of his fellow diners at the Eagle Pub, in Cambridge, UK.[10] He had an important announcement to make: he and an American geneticist named James Watson had "discovered the secret of life." That secret, according to Watson and Crick, was the physical structure of a complex molecule, DNA, that contains the basic genetic material for virtually all life on earth.

On the road to discovering the structure of DNA, Watson and Crick drew on many tools.[11] Perhaps the most iconic of these was a set of glorified Tinkertoys they used to represent various atoms and the electrical bonds between them.[12] These building blocks allowed Watson and Crick to test hypotheses about the feasibility of diverse molecular structures.

In most ways, the structures they built were nothing like molecules. The pieces were hundreds of millions times bigger than atoms, and they were painted various colors, which atoms decidedly are not. Electron structure was represented by sticks poking out of balls at different angles. And yet, by experimenting with these blocks, Watson and Crick managed to extract crucial insights into the real structure of DNA.

The kind of reasoning Watson and Crick did with their building blocks is ubiquitous in the sciences. They built a *model* as an aid to understanding and inference. Models can take many different forms:

physical structures developed in labs, computer programs, mathematical constructions of various sorts. Usually, a model is some sort of simplified or otherwise tractable system that scientists can manipulate and intervene on, to better learn about a messier or more complex system that we ultimately care about.[13] Watson and Crick could not play with the actual structures of molecules, but they could manipulate their building blocks instead and use the resulting structures to learn about the real system.

In fact, we introduced an example of this kind of model in the last chapter—though we did not explicitly label it as such. Bayes' rule, remember, is a formula for how people ought to change their beliefs in light of new evidence. To apply Bayes' rule, we first need to think of our confidence concerning our various beliefs as represented by probabilities—basically, numbers between 0 and 1 that have to satisfy some further conditions. This whole picture, where degrees of belief are numbers that can change via Bayes' rule as we collect evidence, can be thought of as a simplified mathematical model of how humans might really change their minds.

Of course, this model will not capture most real cases of inference perfectly. But it can nonetheless provide insights into what is going on when our beliefs evolve as we learn about the world. It captures the idea that beliefs come in degrees, and it sets out conditions under which those beliefs should change. For instance, if the evidence we have is very likely to occur if our belief is true, we should become more confident in that belief. If our evidence is unlikely when the belief if true, we should become less confident. As we argued in Chapter 1, this insight alone is useful for thinking about issues regarding whether science can ever deliver certainty about anything—and whether we should care.

Bayesian belief updating gives us a model of how *individual* beliefs change. But as we have just seen in the case of methylmercury, science often needs to be understood on the level of a community, not an individual. How do groups of scientists—such as the one Jane Hightower was part of—share knowledge, evidence, and belief? How do they reach consensus? What do these processes tell us about science?

These questions, too, can be studied by developing and examining models. There are many ways to do this, but here, to keep things simple, we focus on just one framework.[14] Where there are other important models to discuss, we do so in the endnotes.

The framework we focus on was introduced in 1998 by economists Venkatesh Bala and Sanjeev Goyal. It is a mathematical model in which individuals learn about their world both by observing it and by listening to their neighbors. About a decade after Bala and Goyal introduced their model, the philosopher of science Kevin Zollman, now at Carnegie Mellon University, used it to represent scientists and their networks of interaction.[15] We use the model, and variations based on it, much as Zollman did.

Why might models be useful here? Communities of scientists are vastly complex. We can investigate them using experiments and case studies, but there are some things that even these powerful methods cannot do for us. For example, we could never track the full progress of an idea, such as that methylmercury was poisoning fish eaters, through an entire scientific network. Where did each scientist first hear of it? When did he or she become convinced it was correct? Who did that scientist share it with? This is especially true of scientific insights that happened in the deep past, and ones that involved large networks of researchers. Models can help fill the gaps in our understanding of how beliefs spread in communities of scientists, and knowledge seekers more generally.

Of course, a model of scientists gathering evidence and communicating with one another cannot capture every detail of how scientific ideas develop and spread. For example, we will not attempt to model the "Eureka moment"—the dawning of that brilliant idea that moves a field forward. (Though, again, we are skeptical that such moments play the significant role history tends to grant them.) Nor will we model power dynamics between scientists, or the role that prestige and timing play in the uptake of scientific ideas.[16] We focus just on the dynamics of belief and evidence.

Even this very simplified model can give us surprising information that we could get no other way. It provides a new way of thinking about how beliefs spread in a community—and a way to ask how those dynamics would change under various conditions.

The basic setup of Bala and Goyal's model is that there is a group of simple *agents*—highly idealized representations of scientists, or knowledge seekers—who are trying to choose between two actions and who use information gathered by themselves and by others to make this choice. The two actions are assumed to differ in how likely they are to yield a desired outcome. This could represent the choice between eating fish or not and so increasing or decreasing one's risk of mercury poisoning; or it could be regulating smokestack emissions and so increasing or decreasing the risks of acid rain. For a very simple example, imagine someone faced with two slot machines, trying to figure out which one pays out more often.[17]

Over a series of rounds, each scientist in the model chooses one action or the other. They make their choices on the basis of what they currently believe about the problem, and they record the results of their actions. To begin with, the scientists are not sure about which action is more likely to yield the desired outcome. But as they make their choices, they gradually see what sorts of outcomes each action yields. These outcomes are the evidence they use to update their beliefs. Importantly, each scientist develops beliefs based not only on the outcomes of their own actions, but also on those of their colleagues and friends.

A clinician like Hightower, for example, might observe what happens to her own patients and also hear about her colleagues' patients. She will use all of these observations in deciding whether she thinks her patients' symptoms are due to mercury poisoning. Similarly, while at the casino you might favor one slot machine, but after hearing from all your friends that they hit jackpots on another, you might change your mind.

In the model, one of the two actions—call it action B—is, in fact, better than action A. (To keep this straight, remember that A is for "All right," but B is for "Better.") But figuring out which action is superior is not necessarily easy. A crucial assumption in this model is that evidence is probabilistic, meaning that when the scientists investigate the world—test a slot machine or warn a sick patient off fish—the results are not always the same. Action B is better than action A because, on average, it yields better results. But there can be many individual instances when action A happens to yield a better result.

In this way, we can think of action B as similar to a biased coin. It may land heads up more often than an ordinary coin—but that does not mean that it never lands tails up. And if you flip a biased coin and an unbiased coin some number of times, there is no guarantee that the biased one will land heads up more often. It is merely *likely* that it will do so.

Not all science looks like this. If you were investigating the laws of gravity, for instance, and you dropped a bowling ball off the top of the Empire State Building again and again, very carefully timing it on each attempt, the results would be remarkably consistent. Likewise for mixing natural gas, oxygen, and a flame: we know what will happen.

But in many types of science, evidence is not so dependable. Again, think of methylmercury. Individual sensitivity to the toxin varies widely, meaning that two people eating the same amounts of swordfish might show very different symptoms. To make matters worse, the symptoms take time to develop. In retrospect, it is easy to look back on Chinese emperors taking mercury tablets to become immortal and think, "How stupid! How did they miss that the stuff is toxic?" But mercury has historically been used again and again in medicine, because without statistical methods it is actually quite difficult to definitively link its use to its harms. The effects are too variable. In cases like this, scientific consensus is hard to reach, and models like the one we are describing can help us understand how that consensus comes about.

We should also emphasize that, although our examples come from science and we are calling the agents in our model "scientists," these models can represent any group of people who are trying to make their way in an unpredictable world. All of us act as scientists sometimes, when we make decisions based on our own experiences and those of our friends. Ever buy a car? There is a good chance that you took it for a test drive and asked the dealer some questions. You were gathering evidence before making a decision. Did you also ask your friends or relatives for advice? Or look at online reviews? If so, you consulted a network of other agents who likewise had gathered evidence, and you used their experiences to influence your beliefs—and ultimately your actions. So these models can apply very broadly. (We will return to this point in Chapter 4.)

We described Bala and Goyal's models as mathematical. At this point, you might wonder where the math comes in. Let us dive into the details a bit more. We have been using anthropomorphic language, talking about "scientists" who "decide" to "act" on the basis of their "beliefs." But in fact we are talking about computer simulations—there are no real decisions here, no physical actions, and no minds that could hold beliefs. Instead, we have an abstract network consisting of a collection of "nodes," each of which may or may not be connected to other nodes by what is called an "edge."

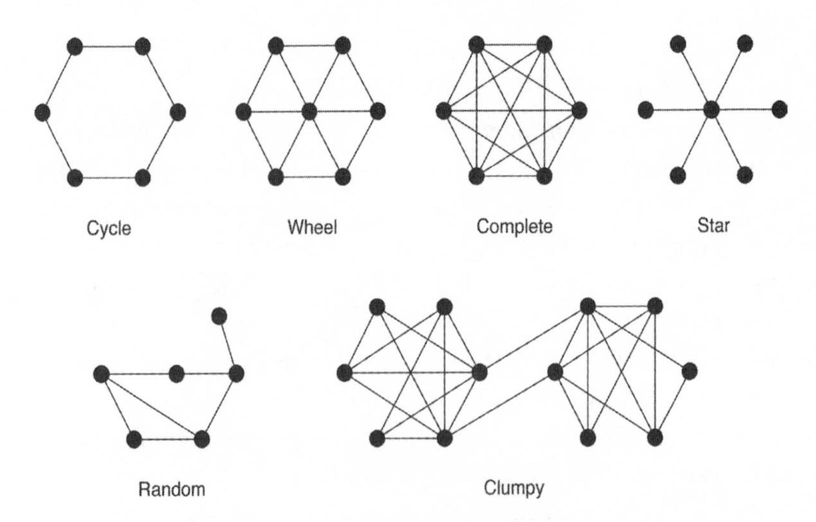

Figure 1. A collection of communication networks. In each network, the nodes represent individuals, or agents, and the connections between them, called edges, represent social ties. Some networks, like the complete, are more densely connected, and others, like the cycle, are more sparse. The clumpy network involves cliques. In the star and wheel networks, some individuals are more central than others. These structures influence how beliefs spread through the network.

Each node represents a scientist, and each edge connects two scientists who have access to each other's results.

These networks can take different shapes. Figure 1 shows some examples of what the communication networks of scientists might look like. Some of these follow patterns: the cycle is a ring with each individual connected to two others; the complete network directly connects all agents to all other agents; and both the star and the wheel have one central node, with the rest of the individuals in the wheel loosely connected and in the star not connected at all. Real human networks are not so neat. They often have substructures that mimic the more regular ones, but they are also "clumpy," with random links between well-connected cliques.[18] As we will see, these structures are often important in determining how information and ideas flow through a group.

How does a node make decisions or take actions? In the model, each node—each scientist—is associated with a number between 0 and 1. This number represents the scientist's level of certainty, or credence, that action B is better than action A. An assignment of .7 would mean that particular scientist thinks there is a 70 percent chance that action B is better than action A. Which action the scientist takes is wholly determined by this number. If it is greater than .5, the scientist performs action B—by which we mean that we simulate pulling a slot machine some number of times and counting the number of times it pays off. Then we use Bayes' rule to update the scientist's credence in light of this result, and likewise update the credences of all of the other neighboring scientists on the network.

If the scientist's belief is less than .5, he or she performs action A. In the simplest version of the model, we assume that everyone knows that this action works exactly half the time.[19] You can think of this as a situation in which, say, a new medical treatment (action B) has been introduced to a market where another well-studied and well-understood treatment (action A) is already available.[20] Doctors are interested only in whether the new treatment is better than the old one; they already know how well the old one works. The fact that we have a network of scientists, however, means that any particular scientist can get evidence of the new treatment's efficacy from their neighbors, even if they do not perform that action themselves. This is like the other physicians who learned of Dr. Hightower's results, even though it never occurred to them to test their own patients' mercury levels.

Figure 2 shows an example of what this process might look like. First, in (a) we see a network of six nodes (scientists) and edges (their connections). Each scientist has a credence ranging from 0 to

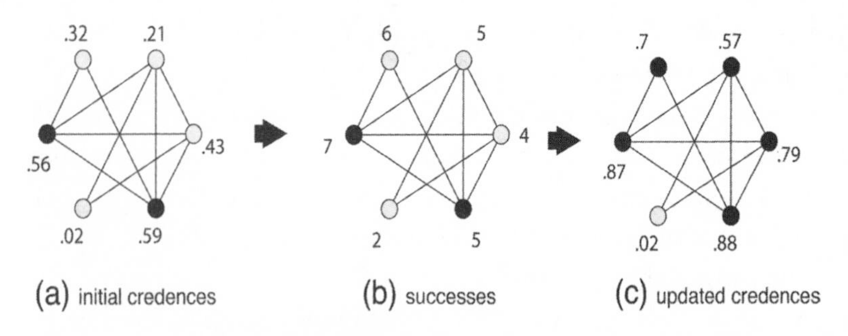

Figure 2. An example of updating and experimentation in a Bala and Goyal–style model. Scientists start with initial credences (a) and use these to decide how they will test the world (b). Light nodes represent those taking action A, and dark nodes, B. In (c) we see that scientists who observe tests of action B update their credences.

1 that action B is better. We can also see that on the basis of their credences in this particular network, four scientists will perform A (the light nodes) and two will perform B (the dark ones). Say they each perform their action ten times. In (b) we can see an example of results they might have obtained (2, 5, 7, etc.) Then in (c) we see how each scientist changes credences using Bayes' rule on the basis of the outcomes observed by themselves and their neighbors. Anyone connected to someone who tried action B—the new, unknown treatment—will update their beliefs. (The scientist with credence .02 does not update since that scientist is not connected to anyone trying action B.) In this case all but one scientist increased their confidence in B, since, as expected, it tended to succeed more often than A. In fact, we can see that when they act next, five scientists will try B instead of A.[21]

This process continues stepwise (try actions, update credences, try actions, update credences) until the scientists have converged on a consensus. This can happen if all of the scientists have sufficiently high credence—greater than .99—that action B is better; or all of them have sufficiently low credence, less than .5, so that no one in the network ever performs action B, in which case they will not learn anything further about it. In the first case, we say the network has converged to the true belief. In the second, we say it has converged to the false one. In general, these models tend to converge to the true consensus—that is, the whole network comes to believe that action B is better. But, as we will see, they sometimes go to the false one.[22]

What we want to understand is this: Under what circumstances do networks of scientists converge to false beliefs?

Stomach ulcers are painful sores in the lining of the stomach. It turns out that they are caused by a kind of bacteria known as *H. pylori*.[23] Decisively showing that bacteria cause ulcers ultimately earned the 2005 Nobel Prize for two Australian medical researchers, Robin Warren and Barry Marshall, who managed to convince their fellow scientists of this relationship during the 1980s. But it is a bit strange to say that Warren and Marshall *discovered* the link. In fact, the theory that ulcers were caused by bacteria dates back to 1874, when a German bacteriologist by the name of Böttcher and a French collaborator, Letulle, isolated bacterial colonies in an ulcer and argued that the bacteria were the ulcer's cause.[24] During the following decades, evidence slowly accumulated that bacteria were, indeed, responsible for ulcers.

But the bacterial theory was not the only one available. The other possibility, also accepted by many doctors and scientists, was that stomach acid was the culprit. In the early twentieth century, scientists investigated both theories and found evidence in favor of each. But then, in 1954, the bacterial theory suffered a devastating setback. Gastroenterologist E. D. Palmer biopsied the stomachs of more than one thousand patients and found no evidence of bacteria at all.[25] The conclusion seemed to be that bacteria could not live in the human stomach, meaning that they could not possibly cause ulcers.

Palmer's results essentially ended attempts to confirm the bacterial theory—aside from a few isolated doctors who continued to successfully treat ulcer patients with antibiotics. (Inhibiting gastric acid also helped—though ulcers treated in this way tended to return.) It was not until almost thirty years after Palmer published his results, when Warren observed a new strain of bacteria in stomach biopsies taken near tissue with ulcers, that serious research on the bacterial theory picked up steam again. Later, Marshall managed to isolate and cultivate the new strain, showing definitively that bacteria *could* live in the human stomach after all.

Even with these strong results, Warren and Marshall faced significant skepticism. The acid theory was widely held and deeply ingrained. The resistance was so strong that Marshall resorted to dramatic stunts to attract attention—and adherents—to their theory. In a fit of pique, he apparently drank a petri dish full of *H. pylori* himself and then successfully treated the ensuing ulcer with antibiotics.[26] Ultimately, Warren and Marshall managed to persuade their colleagues that the bacterial theory was right. But this episode could very well have gone differently. Had there not been a few scientists willing to give the bacterial theory a chance, we might still be using antacids to treat recurring ulcers.

How could this happen? One of the most startling findings from the Bala-Goyal models is just how strongly people's beliefs can influence one another. If we imagined a group of agents with no network connections gathering probabilistic evidence (and not sharing it), we would expect some of them to end up with the right theory and some with the wrong one. For instance, scientists who play the better slot machine and happen to lose all their money may give up on that machine for good. But with no communication, we should not expect much correlation between the various scientists' beliefs. Some scientists would have good luck and stick with the better machine; others would not. Each conclusion would be independent from all the others.

Once scientists start to share evidence, however, it becomes extremely likely that they will all come to believe the same thing, for better or worse.[27] Notice that this happens in the models only because the scientists share evidence. There is no psychology here. No one is imitating anyone else, no one is trying to conform, no one is smarter or dumber than the others. There are no thought leaders or sheeple.

Why does it happen? Imagine a group of scientists gathering and sharing data. Suppose a few of them try the better action—reducing fish consumption, say, on the hypothesis that eating too much fish can cause mercury poisoning. As they continue to gather evidence, it starts to influence their colleagues and neighbors, just as we saw in the Hightower case. Some of these come to believe the right theory and now start to gather evidence about it themselves. They, in turn, can persuade new colleagues and neighbors. The belief spreads throughout the network until everyone agrees.

Notably, this means that a successful *new* belief can spread in a way that would not have been very likely without the ability to share evidence. Suppose that almost every scientist starts with an extant belief (say, the mercury in fish is not poisoning people). We do not expect them to gather evidence about mercury and fish—why would they? Without data sharing, the chance that each independently decides to test this new possibility is miniscule. With data sharing, however, it takes just one scientist to start testing a new hypothesis for it to start catching hold throughout the scientific network (if the scientist gets positive results).

Figure 3 shows what this might look like. It is a simplified image (showing just the updating of credences, but not the successes) of

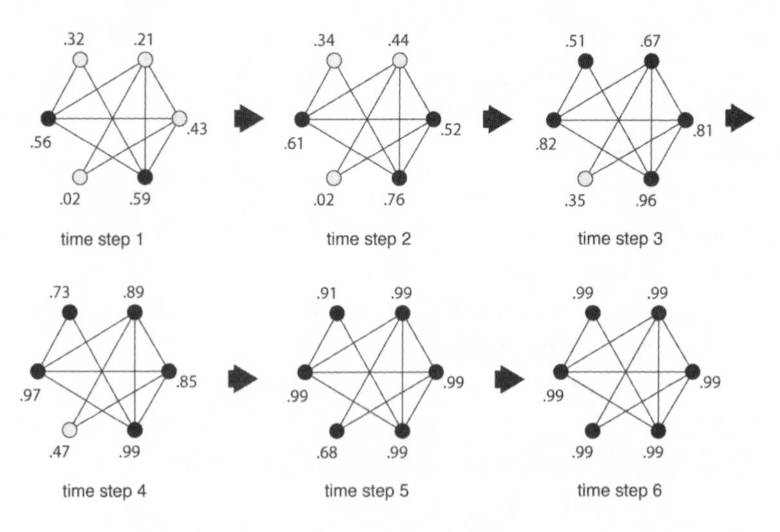

Figure 3. An example of a network that achieves convergence on true beliefs. Light nodes represent belief in A and dark nodes belief in B. In each time step agents are testing their beliefs and updating their credences on the basis of their results and their neighbors' results. As time goes on, more agents have high credences in the true belief until the entire network becomes essentially certain that action B is better.

the process like that shown in figure 2. In each subsequent round, more scientists are persuaded by the results of their neighbors to try the better action, and eventually it spreads throughout the network.

This is the optimistic outcome. As we argued in the Introduction, the social spread of knowledge is a double-edged sword. It gives us remarkable capabilities, as a species, to develop sophisticated knowledge about the world, but it also opens the door to the spread of false belief. We see this in the models as well: especially when scientists tackle hard problems, they can all come to agree on the wrong thing. This happens when a few scientists get a string of misleading results and share them with their colleagues. Scientists who might have been on track to believe the true thing can be derailed by their peers' misleading evidence. When this happens, the scientists would have been better off *not* getting input from others.

It is worth taking a moment to let this sink in. Usually, when scientists behave rationally but gather uncertain data, sharing evidence helps the whole group get to the right belief, even persuading those who were initially skeptical. But sometimes this process backfires, and communication between scientists actually leads to a consensus around the false belief. Remember the Vegetable Lamb. Without communication among learned scholars, this bizarre belief would never have gone anywhere. The sharing of evidence ("I tasted its wondirfulle flesh!") convinced many with correct beliefs that the wrong thing was true.

This trade-off, where connections propagate true beliefs but also open channels for the spread of misleading evidence, means that sometimes it is actually better for a group of scientists to communicate less, especially when they work on a hard problem. This phenomenon, in which scientists improve their beliefs by failing to communicate, is known as the "Zollman effect," after Kevin Zollman, who discovered it.[28] If everybody shares evidence, a chance string of bad data can persuade the entire group to abandon the correct theory. But in a group where not everyone listens to everyone else, pockets of scientists can be protected from misleading data and continue to gather evidence on the true belief that eventually persuades the rest of the community.[29]

Another way to put this is that some temporary diversity of beliefs is crucial for a scientific community. If everyone starts out believing the same thing, they can fail to try out better options. It is important for at least a few people to test different possibilities so that the group will eventually find the best one. One way to maintain this diversity of beliefs for a long enough time is to limit communication, so that researchers' beliefs do not influence one another too much while they test different theories.[30]

As Zollman himself points out, the Zollman effect can help explain how Palmer's results finding no bacteria in the stomach had such a dramatic effect—and why the medical establishment held fast to a false theory for so long.[31] Physicians were tightly connected to one another, so a single result—even though it turned out to be misleading—convinced nearly all of the gastroenterologists in the world that they should abandon what turned out to be the true theory of ulcers. Taking the actions they did was very likely the rational thing to do given Palmer's evidence, which seems to have been very strong. But the structure of the community meant that rational actions by every individual actually made the false belief persist. Had fewer scientists known about Palmer's results, the bacterial theory might have won out sooner.

Of course, Warren and Marshall did, eventually, return to the bacterial theory. If we add to the model the fact that scientists sometimes test the alternative theory—they sporadically or accidentally perform action B, even though they generally do not expect it to be better—they can overcome the Zollman effect, much as they would if they were less tightly connected with one another. But it can be a slow process and relies on luck. On the other hand, it works precisely because of evidence sharing: if strong evidence for a surprising new theory appears in this random way, the connections between scientists will allow the better theory to eventually take hold and spread.[32]

Polly Murray was suffering from fatigue, terrible headaches, and joint pain so severe that she struggled to move.[33] She had seen doctors, but none of them had managed to help her. Many, in fact, hinted that her symptoms might by psychosomatic—or to put it more bluntly, they thought she was a nut. But as Murray meticulously documented, she was not the only one with these symptoms.

Many of her friends and their children, all living in the small town of Lyme, Connecticut, suffered from the same strange cluster of ailments. Two of her children had been diagnosed with juvenile rheumatoid arthritis. This is a rare disease and it is not infectious—it seemed exceedingly unlikely that there could be an epidemic of it.

In 1975, Connecticut health officials took Murray's case to Allen Steere, a rheumatologist working on a fellowship at Yale. Steere met with Murray, and she showed him her list of neighbors with the same symptoms.[34]

Steere's extensive investigation into the possible causes of the ailment eventually yielded a diagnosis: a new tick-borne illness later named Lyme disease, after the town where Murray and her friends lived.[35] A few years later, the strain of bacteria responsible was isolated and named *Borrelia burgdorferi* (after Willy Burgdorfer, who did the isolating).[36] This discovery had a massive impact on patients like Polly Murray. After treatment with antibiotics, many of them regained lives previously lost to debilitating pain. September 24 was declared "Allen Steere Day" in Connecticut to celebrate his findings.

Fast forward twenty-five years. Allen Steere was receiving death threats and hate mail from Lyme patients across the country. Security guards had to be hired to protect him at public appearances. The New England Medical Center, where he was now chief of rheumatology, employed an expert who spent hours each week monitoring the public threat to his safety.

What had happened?

Lyme disease is caused by a spirochete—a type of bacteria shaped like a spiral or helix, like those that cause syphilis. And like syphilis, the disease proceeds in stages. Initial infection causes flulike symptoms: fever, headache, joint aches, and often, but not always, a distinctive rash in the shape of a bull's-eye.[37] As the spirochetes spread throughout the body, some patients develop more alarming symptoms: meningitis, encephalitis, facial paralysis, and mental disturbances.[38]

As with any infection, the human immune system responds by attacking the invader, producing antibodies that help it identify and root out the Lyme spirochete. In many cases, though, this is not enough to totally suppress the infection. *Borrelia* uses its distinctive shape to wriggle into tissues throughout the body, and it employs a host of nasty tricks to hide from the immune system. When left untreated, late-stage Lyme causes the sorts of symptoms that first brought Polly Murray to see Allen Steere: crippling joint pain, numbness and pain in the extremities, brain fog, insomnia, extreme fatigue, and maladies such as serious cognitive impairments.[39]

This much, at least, is relatively uncontroversial. But what happens after Lyme is treated by antibiotics? This question is at the heart of what has become known as the "Lyme wars." It is the Lyme wars that put Allen Steere's safety at risk.

On one side are those who hold the view, widespread within the medical establishment and endorsed by groups such as the US Centers for Disease Control and Prevention (CDC), that a single round of treatment with antibiotics is generally enough to eliminate the Lyme spirochete, and so to cure a patient of the disease.[40] On the other side are a large number of Lyme disease patients who have already undergone antibiotic treatment but who continue to experience debilitating symptoms typical of the disease. On the basis of their experiences, some "Lyme-literate" doctors have developed treatment programs for "chronic Lyme disease," usually involving repeated rounds of heavy antibiotic use.

In the early 1990s, observing the emergence of the Lyme-literate doctor movement, Steere grew concerned that the diagnosis of Lyme disease had become a catchall for other diseases such as fibromyalgia and chronic fatigue syndrome (themselves both poorly understood and controversial). After investigating patients referred to him for Lyme, he formed the opinion that many did not have the disease. Knowing that long-term antibiotic use has serious side effects, he began to publicly advocate for greater caution in Lyme diagnosis and treatment.

Thus began a decades-long battle (which is still raging) over chronic Lyme disease. Steere, and most professional doctors' groups and disease control centers, contend that chronic Lyme is actually a combination of other diseases, plus, perhaps, a mysterious post-Lyme syndrome that might involve a continued immune response to Lyme after it has already been treated. They argue that long-term antibiotic treatments do serious harm to sick patients, without any benefits.[41] Most chronic Lyme patients, they point out, do not test positive for the Lyme spirochete, and four large studies conducted by the National Institutes of Health have each shown that long-term antibiotic treatments do not improve these patients' symptoms.[42]

On the other side of the debate are the patients, Lyme-literate physicians, and various advocacy organizations. They contend that Lyme spirochetes often hide in the body, avoiding total eradication by standard antibiotic treatments, and that long-term antibiotics are an effective treatment.[43] The doctors involved claim to have successfully treated thousands of patients. They refer to evidence showing that Lyme can survive aggressive antibiotic treatment in dogs, mice, and monkeys[44] and can subsequently reinfect ticks and other hosts with live spirochetes despite sometimes failing to show up in standard tests.[45] Disaffected by what they see as a wall of opposition from mainstream researchers, the Lyme Disease Foundation even started its own publication—the *Journal of Spirochetal and Tick-Borne Diseases*—to publish results defending the existence of chronic Lyme.

At stake in this debate is the well-being of thousands of suffering patients. They vilify Steere and others who maintain that those with chronic Lyme should not be treated indefinitely. Some argue that these doctors are conspiring to hide the truth about chronic Lyme—possibly because they are in cahoots with insurance companies that do not want to pay for long-term treatment.[46]

If threats of violence against doctors seem extreme, note that the medical establishment has its own weapons. Patients are denied insurance coverage for expensive treatments that they claim reduce their symptoms. Doctors willing to prescribe long-term antibiotic treatments are often regarded as quacks and pariahs by their colleagues and by medical licensing boards. Some of the most prominent of these doctors, such as the beloved Charles Ray Jones, who has treated thousands of children for chronic Lyme, have been disciplined by licensing boards or had their licenses suspended.[47]

On both sides, the Lyme wars have extended far beyond discussions over coffee at academic conferences and in the pages of medical journals. And one side is putting people's lives at risk. The only question is which.

On June 14, 2017, in Alexandria, Virginia, a group of Republican congressional representatives met to practice for the Congressional Baseball Game for Charity, which was scheduled for the following day.[48] Suddenly, mid-practice, shots rang out from near the third-base dugout. Congressman Steve Scalise was hit in the hip; a lobbyist, a congressional aide, and a police officer assigned to protect Scalise were also shot and injured. The gunman was shot and died of his wounds.

The shots were fired by a left-wing extremist named James Thomas Hodgkinson. Hodgkinson reportedly belonged to Facebook groups with names like "The Road to Hell Is Paved with Republicans," where he posted vitriolic anti-Trump comments daily.[49]

Two months later, white supremacists, neo-Nazis, and other nationalist and nativist extremists marched through Charlottesville,

Virginia, carrying torches and chanting anti-Semitic, racist, and pro-Trump slogans.[50] Violence broke out between the "Unite the Right" crowd and counterprotesters, injuring fourteen people. The following day, a twenty-year-old white supremacist named James Alex Fields Jr. drove his car into a counterprotest. He injured nineteen people and killed a thirty-two-year-old woman named Heather Heyer. In the month before the attack he reportedly had posted photos of Nazis, swastikas, and pro-Trump memes on his Facebook page, as well as pictures of alt-right icons such as Pepe the Frog.[51]

The term "polarization" originated in physics to describe the way some electromagnetic waves propagate in two oppositely oriented ways. By the mid-nineteenth century, political pundits had embraced this metaphor, of two opposite ways of being, to describe disagreements in a state dominated by two parties. Today it captures the broad sense that Democrats and Republicans, Labour and Tories, left-wing and right-wing, are increasingly divided in their beliefs and moral stances.

Hallmarks of polarization include individuals on two sides of an issue who tend to move farther from consensus, rather than closer to it, as debate progresses. In some instances of political polarization, moral mistrust breeds between those who disagree, sometimes leading to violence, as in the shooting of Steve Scalise and the killing of Heather Heyer.

In the case of chronic Lyme disease, we see a situation where a *scientific* community has polarized over a set of scientific beliefs in much the way that some communities polarize over political beliefs. Here, too, the situation has progressed to threats of violence.

This situation may seem surprising. We tend to think of political stances and scientific beliefs as importantly different. Political stances are motivated by social values: moral norms, religious beliefs, and beliefs about social and economic justice. We adopt political positions because we want to promote something we value in our country and our lives. Scientific beliefs, on the other hand, are supposed to be value-free (arguments from Chapter 1 notwithstanding). In an ideal science, thinkers adopt beliefs that are supported by evidence, regardless of their social consequences.

In fact, this is not how science works. Scientists are people; like anyone else, they care about their communities, their friends, and their country. They have religious and political beliefs. They value their jobs, their economic standing, and their professional status. And these values come into play in determining which beliefs they support and which theories they adopt.[52]

That said, it is not clear, in the case of the polarization over Lyme disease, that differing values play much of a role. The physicians on both sides of the debate seem to have the same values. Allen Steere has devoted his professional life to studying and treating the disease. His objections to patients taking heavy doses of antibiotics seem to be genuinely motivated by concern for their health and safety. At the same time, doctors such as Charles Ray Jones are trying to treat patients who are truly suffering, and, on their own reports, they are succeeding in doing so. Everybody involved wants to protect and cure the afflicted.[53]

Besides having the same values, the two sides in the chronic Lyme case have access, for the most part, to the same evidence. They can, and often do, read the same journal articles about Lyme disease. They see patients with similar symptoms. Inasmuch as Lyme-literate physicians prescribe long-term antibiotics and most other physicians do not, these groups will not always observe patients undergoing the same sorts of treatments, but all of them read the same reports of randomized controlled trials on the effects of antibiotic treatments, and they can discuss other doctors' clinical observations.

So how have things gotten so polarized? The models of scientific networks we have described in this chapter suggest that scientific communities should tend strongly toward consensus as they gather and share evidence. Eventually, influence and data flowing between researchers should sway the whole group one way or another.

Or perhaps not. The models we have considered so far assume that all scientists treat all evidence the same way, irrespective of the source. But is that reasonable? Do all scientists trust one another equally? Do they consider all other researchers equally reliable?

Consider a small alteration to the model we introduced earlier. Suppose scientists in a network do not treat all the evidence the same way but instead take into account how much they trust the colleague who is sharing research with them. This is hardly an unreasonable thing to do. It is, in fact, an essential part of science—and scientific training—to evaluate the quality of the evidence one encounters, and to exercise judgment in reacting to putative evidence. Taking into account the source of reported data is surely a natural way to do this. Scientists who rely on studies written by known quacks are arguably abdicating their responsibilities.

How can we include this sort of "trust" in the Bala-Goyal model? Here is one suggestion. Suppose scientists tend to place greater trust in colleagues who have reached the same conclusions they have reached, and less in those who hold radically different beliefs. Again, this is not so unreasonable. We all tend to think we are good at evaluating evidence; it is only reasonable to think that those investigating similar problems, who have reached different conclusions, must not be doing it very well.[54]

We can thus change how the scientists in our model update their beliefs in light of new evidence. The rule we have used so far, Bayes' rule, takes for granted that we are certain that the evidence we are considering was really observed: there were no errors, no subterfuge, no miscommunications. This is a highly idealized case. Usually, when we encounter evidence, it is not perfectly certain. In such cases, there is a different rule that can be used to update your beliefs, called "Jeffrey's rule," after Princeton philosopher Dick Jeffrey, who proposed it. Jeffrey's rule takes into account an agent's degree of uncertainty about some piece of evidence when determining what the agent's new credence should be.[55]

But how much uncertainty should the scientists assign to any particular piece of evidence? Suppose they do this by looking at how far the other scientists' beliefs are from their own, and letting that distance determine their degree of uncertainty. Reading Allen Steere's newest article, a Lyme-literate physician does not fully trust the reported results. Hearing about the clinical experiences of Charles Ray Jones, an establishment researcher is skeptical. In one version of this model, the scientists simply stop listening at some point and do not update their beliefs at all on the basis of evidence produced by someone who disagrees with them too much. In another version, the scientists could think that the scientists who disagree too much are corrupt or otherwise trying to mislead them and therefore assume that the evidence they have shared is actively fabricated. In this case, they would update their beliefs in the other direction.[56]

This small change to the model radically alters the outcomes. Now, instead of steadily trending toward a consensus, either right or wrong, scientists regularly split into polarized groups holding different beliefs, with each side trusting the evidence of only those who already agree with them.[57] Initially, scientists' beliefs are randomly distributed throughout the network. Most scientists begin by listening to, and updating on the basis of, the evidence produced by most other scientists. But over time, groups of scientists begin to pull apart until eventually you have two groups with opposite beliefs who do not listen to each other at all.

Such a model does not capture the moral anger we see in the case of chronic Lyme, or in political polarization. But we do see that under fairly minimal assumptions, entire scientific communities can

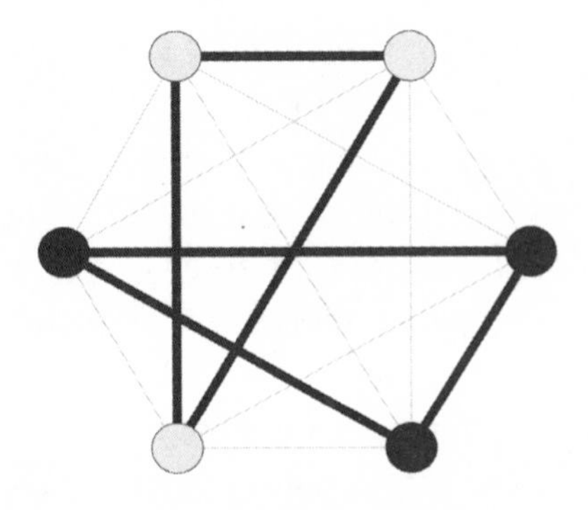

Figure 4. A complete, that is, fully connected, network in which agents are polarized in that they have stable, opposing beliefs. Light nodes represent those taking action A, and dark nodes, B. The weights of the connections between the nodes represent trust between agents—which translates into belief that other agents share real data. Within each group, agents trust others' data, but they do not trust data from the other group.

split into two groups with opposite beliefs. Even worse, this sort of polarization is stable: no amount of evidence from the scientists who have adopted the correct belief will be enough to convince those who adopted the wrong belief. And the polarization does not depend on individuals not seeing the evidence of those with different beliefs. They receive this evidence just as before. They simply do not believe it.

Figure 4 shows a network in which all people see each other's evidence (a complete network) but that has moved toward polarization. The shade of the nodes represents which belief each individual espouses (light for A and dark for B), and the weight of each connection represents the degree of trust the agents give to each other's evidence. As you can see, there are two groups with opposing beliefs who do not listen to each other.

We also find that the greater the distrust between those with different beliefs, the larger the fraction of the scientific community that eventually ends up with false beliefs. This happens because those who are skeptical of the better theory are precisely those who do not trust those who test it. As this skepticism increases, more agents will fail to update their beliefs in light of new studies pointing toward more accurate beliefs. We can think of this sort of polarization as a way that communication closes down between opposed groups over time. The group holding false beliefs thus becomes insensitive to results pointing to better ones.

These results follow from one way of thinking about how scientists might distrust each other. But there are other possibilities. In a less dramatic version of the model, scientists would listen to everyone but discount the evidence of those who disagree with them rather than ignoring it completely. In models with this assumption, we find that all scientific communities eventually do reach a consensus, just as in the original Bala-Goyal models. But mutual mistrust slows the process dramatically. Even in cases where scientists listen to each other enough that they do not reach stable, polarized outcomes, mistrust among those with different beliefs can produce *transient* polarization—long periods during which some scientists prefer the worse theory and mostly discount the evidence of those who prefer the better one.

One of the more surprising aspects of this transient polarization is that people who start off holding similar positions can end up on opposite sides of a debate. Imagine, for example, that Sally and Joe are scientists, and Sally is initially a bit more skeptical than Joe about a new theory. If Joe gathers evidence supporting the theory, his credence will increase. Sally's credence will also go up, but not as much, because she trusts Joe's data less than Joe does. This means that both their credences are higher than before, but also farther apart. Now Joe gathers more evidence, and his beliefs again jump up. Sally is also more convinced, but since the distance between her and Joe is even greater now, her credence changes even less than the first time.

Eventually, Sally may conclude that Joe's theory is better, but she will take a long time to get there. Or else Joe may approach certainty so much faster than Sally that he leaves her behind. From her perspective, it will look like he is going down a rabbit hole, and she will conclude that he is too radical to trust.

Of course, even transient polarization can be a damaging outcome. In cases like Lyme disease, dire consequences are associated with the wrong belief: either overtreatment with antibiotics or ignoring a dangerous chronic infection. A significant slowing of the emergence of scientific consensus can seriously affect the lives of those with the disease.

Polarization has been studied in many disciplines. There is a large literature, for instance, looking for explanations of polarization in individual psychology. But researchers in this field tend to assume that when two actors look at the same evidence, if they fail to change their beliefs in the same way, then at least one of them must be irrational.[58] After all, you might think, the evidence either supports a given belief or it does not.

For example, many psychologists have shown that people tend to search out and pay attention to only the evidence that accords with their current beliefs. This is known as "confirmation bias"—reasoning by which we tend to confirm our current beliefs—and it is a variety of what is sometimes called "motivated reasoning." A typical psychological experiment on polarization might give participants two sets of evidence, or arguments, for and against an issue, and see how they change their beliefs. Political scientists Charles Taber, Damon Cann, and Simona Kucsova, for example, presented subjects with conflicting evidence on issues ranging from the legalization of marijuana to the Electoral College. They found that those who started with strong beliefs about these issues became only more entrenched during the study—irrespective of what their starting beliefs were or what evidence they were given.[59] The proposed explanation is that the subjects paid attention only to evidence supporting the view they already held.

We are not suggesting that this psychological effect does not occur. It seems it does—and it is very likely a factor in real-world polarization. But the models of polarization based on Jeffrey's rule that we have described strongly suggest that psychological biases are not necessary for polarization to result. Notice that our agents do not engage in confirmation bias at all—they update on any evidence that comes from a trusted source. Even if people behave very reasonably upon receiving evidence from their peers, they can still end up at odds.

These models can inform our understanding of political polarization as well as the polarization of a scientific group. Sometimes, polarization happens over a moral/social position. The abortion debate, for instance, is obviously extremely contentious, and most of the debate is not over facts but over whether it is inexcusably wrong to abort unwanted fetuses.

But in other cases, we see political polarization arise over matters of scientific fact. When it comes to climate change, for instance, the debate is not primarily about whether something is morally right or wrong, or whether an economic policy is just or not. Rather, the disagreement seems to be about whether carbon emissions from human sources actually contribute to changes in weather patterns. This is not a matter of morality or values: either greenhouse gases are affecting the climate, or they are not.

Of course, there is little question that industrial interests have obscured the scientific consensus on the causes of climate change, by spreading misinformation and creating a sense of controversy. But the models we have discussed suggest that even without industrial interference, a community of people trying to choose scientific beliefs to guide their votes or policy choices can end up with this sort of disagreement.

The take-away is that if we want to develop successful scientific theories to help us anticipate the consequences of our choices, mistrusting those with different beliefs is toxic. It can create polarized camps that fail to listen to the real, trustworthy evidence coming from the opposite side. In general, it means that a smaller proportion of the community ultimately arrives at true beliefs.

Of course, the opposite can also happen: sometimes, too much trust can lead you astray, especially when agents in a community have strong incentives to convince you of a particular view. The models we have considered so far assume that all scientists accurately report their results. In this sort of case, it makes little sense to discount the results of those you disagree with. But this is not the universal case. In fact, in the next chapter, drawing on the modeling work of philosopher Bennett Holman at Yonsei University and philosopher and political scientist Justin Bruner at the Australian National University, we discuss how important discounting the evidence of others can be when industry attempts to influence science.

Ultimately, as we will see, when assessing evidence from others, it is best to judge it on its own merits, rather than on the beliefs of those who present it.

In 1846 Ignaz Semmelweis, a Hungarian physician, took a post in the first obstetrical clinic of the Vienna General Hospital. He soon noticed a troubling pattern. The hospital's two clinics provided free care for poor women if they were willing to be treated by students—doctors in the first clinic, where Semmelweis was stationed, and midwives in the second. But things were not going well in the first clinic.[60]

Puerperal, or "childbed" fever, was rampant, killing 10 percent of patients on average. Meanwhile, in the second clinic, the presumably less knowledgeable midwives were losing only 3–4 percent of their patients. Even more surprising, the death rate for women who had so-called street births on the way to the hospital was much lower than for women who received the dubious help of the doctors-in-training. During Semmelweis's first year, the first clinic's reputation was so bad that his patients literally begged on their knees to be transferred to the second.

Dismayed by their record, and horrified by the terrible deaths his patients were enduring, Semmelweis set out to find the cause of the clinic's high fever rates. In March 1847, he had a breakthrough. A colleague died of symptoms very similar to childbed fever after receiving a small accidental cut during an autopsy. Semmelweis connected this incident with the fact that obstetricians in the first clinic regularly attended patients immediately after conducting autopsies on diseased corpses. Childbed fever, he concluded, was a result of "cadaverous particles" transferred via the student doctors' hands. After he started requiring regular hand-washing with a chlorinated solution, the clinic's death rate plummeted.

Toward the end of 1847, Semmelweis and his students published their findings in several prominent medical journals.[61] He believed his innovation would revolutionize medical practice and save the lives of countless women. But instead, his fellow physicians—principally upper-class gentlemen—were offended by the implication that their hands were unclean, and they questioned the scientific basis of his "cadaverous particles," which did not accord with their theories of disease. Shortly thereafter Semmelweis was replaced at the Vienna General Hospital. In his new position, at a small hospital in Budapest, his methods brought the death rate from childbed fever down to less than 1 percent.

Over the remaining eighteen years of his life, Semmelweis's revolutionary techniques languished. He grew increasingly frustrated with the medical establishment and eventually suffered a nervous breakdown. He was beaten by guards at a Viennese mental hospital and died of blood poisoning two weeks later, at the age of forty-two.

Semmelweis was right about the connection between autopsies and puerperal fever, and the decisions he made on this basis had meaningful consequences. He saved the lives of thousands of infants and women. But his ideas could have saved many more lives if he had been able to convince others of what he knew. In this case, although he communicated his beliefs to other scientists and provided as much evidence as they could possibly desire, his ideas were still rejected, at great cost. The warrantless belief persisted that gentlemen could not communicate disease via contact.

The puzzling thing about the Semmelweis case is that the evidence was very strong. In this way, the case was not like the Vegetable Lamb—or even like Lyme disease, mercury poisoning, or the other relatively difficult cases we have discussed. The message from the world was loud and clear: hand-washing dramatically reduces death by puerperal fever. What went wrong?

On January 21, 2017, Donald Trump was inaugurated as the forty-fifth president of the United States. Within hours, his new administration was engulfed in a bizarre media storm. The subject had nothing to do with policy or foreign affairs. The brouhaha was over the size of Trump's inauguration crowd.

In his first White House press conference, Sean Spicer, then the Trump Administration press secretary, declared that Trump had had the "largest audience to ever witness an inauguration." But counts of Washington, D.C., Metro ridership and crowd estimates based on overhead photographs seemed to definitively show that Spicer's claim was bogus.[62] Several media outlets reported that the attendance was underwhelming compared with the previous two inaugurations of Barack Obama, let alone the massive Women's March in Washington protesting the administration, which occurred the following day. Very quickly the White House's denial of basic facts became the story.[63]

Political scientist Brian Schaffner and pollster Samantha Luks conducted a study of this incident.[64] They showed almost fourteen hundred American adults photos of two inauguration crowds side by side. They then asked half of the survey participants which photo was from Trump's inauguration and which from Obama's. Perhaps unsurprisingly, more Trump supporters than Hillary Clinton supporters falsely identified the more crowded photo as from the Trump inauguration.

The other half of study participants were asked a different, presumably easier question: which photo had more people? The shocking result was that 15 percent of Trump supporters chose the photo with the clearly smaller crowd. They ignored the stark evidence in front of them and agreed with Spicer. Schaffner and Luks interpreted these results as evidence that the respondents wished to signal their strong support for their candidate. But there is another explanation that draws on a large literature in psychology, concerning a phenomenon known as "conformity bias."

In 1951, a psychologist at Swarthmore College named Solomon Asch devised a now-classic experiment.[65] He showed groups of eight participants a card with one line on the left and three lines of varying length on the right (figure 5). Their task was to identify which line on the right was as long as the line on the left. Unbeknownst to his subjects, seven members of the group were confederates who

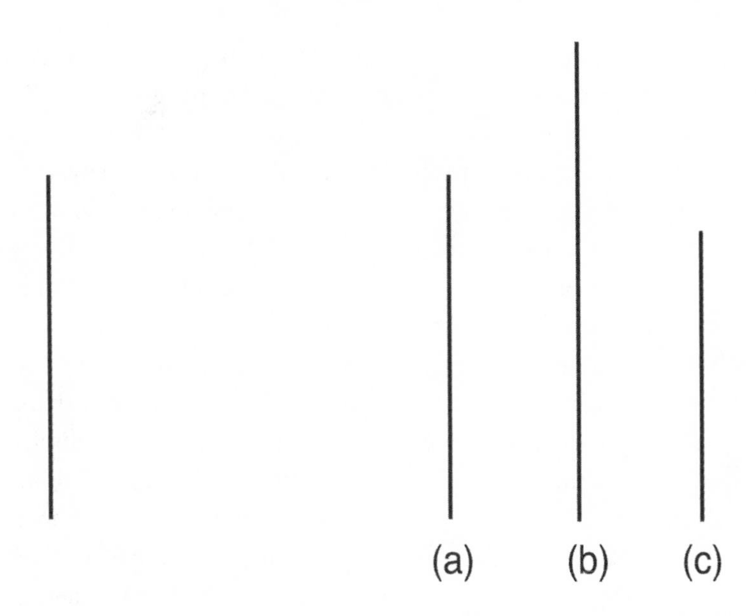

Figure 5. The prompt for Solomon Asch's conformity experiment. Even though (a) matches the length of the line on the left, confederates would all choose the same, incorrect line (b or c). Subjects then had to choose whether to conform or to choose the correct line (a).

were all instructed to choose the same wrong line. For example, in figure 5, they might all, incorrectly, choose *c* instead of *a*. The subject, made to answer last, then had a choice: he or she could either agree with the rest of the group and pick *c*, or pick the correct line *a*. More than a third of study participants agreed with the others in the group. They chose to go against the evidence of their own senses in order to conform with what the others in the group did.

While conformity seems to vary across cultures and over time, it reflects two truths about human psychology: we do not like to disagree with others, and we often trust the judgments of others over our own.[66]

Notice that sometimes, the latter is a reasonable strategy. No one is perfect, and so it can be a good idea to doubt ourselves in the face of disagreement, especially if many others seem to independently agree. In fact, a classic mathematical theorem, due to the eighteenth-century French mathematician Marquis de Condorcet, covers a case very close to this.[67] Suppose you have a group of people who are trying to make a judgment about something where there are two possible answers and only one of them is correct. If each person is individually more likely than not to get the correct answer, the probability that the whole group will get the right answer by voting increases as you add more and more voters. This suggests that there are cases when it is actually a good idea to accept your own fallibility and go with the majority opinion: by aggregating many fallible voices, you increase the chances of getting the right answer.[68] Those who have watched the game show *Who Wants to Be a Millionaire?* will be familiar with this effect. Contestants who poll the audience for the answer to a question can expect correct feedback 91 percent of the time, compared with those who ask a single friend and get the right answer 65 percent of the time.[69]

But trusting the judgments of others does not always work so neatly when these judgments are not actually independent of each other. UCLA economists Sushil Bikhchandani, David Hirshleifer, and Ivo Welch, for instance, have described a phenomenon known as an "information cascade," by which a belief can spread through a group despite the presence of strong evidence to the contrary.[70] In these cases, incorrect statements of belief can snowball as people's judgments are influenced by others in their social environment.

To see how this works, imagine a group in which every member has private information, just as in the Condorcet case, that they can use to make a judgment call. Suppose, for instance, that each person has a tip—usually dependable, but sometimes wrong—about which of two stocks will perform better during the next month. Say that in a group of fifty people, forty-eight have secret information suggesting that Nissan will perform best, and just two have information suggesting that General Motors stock will outperform Nissan. Suppose that those two people publicly buy GM stock. The next investor sees this and, perfectly reasonably, infers that the first two people had private reasons to think GM was preferable. On these grounds, that investor might conclude that his or her own private evidence for Nissan is not as strong as the overall evidence for GM. So that investor decides to buy GM.

Now any other investor will see that three people have purchased GM stock, presumably on the basis of their private information. This will give them an even stronger reason to think that GM is preferable to Nissan, contrary to their own private information. Soon everyone is buying GM, even though almost no one independently would think it was a good idea.

Like other models we have looked at, models of information cascades reveal that individuals acting rationally—making the best judgments they can on the basis of the available evidence and their inferences about others' beliefs based on behavior—can fall into a trap. A group in which almost every member individually would be inclined to make the right judgment might end up agreeing collectively on the wrong one.

Information cascades are not the same as conformity bias; the individuals in the stock-trading case are not trying to fit in with the group. They are making rational decisions on the basis of the evidence available to them, which includes both their own private information and the actions of others. We do not actually think information cascading explains the behaviors of Trump supporters, or of the doctors who ignored Semmelweis. The point is simply that even in a case where conforming might seem like a generally good thing—because others might have information we lack—the whole group can end up behaving in a highly irrational way when our actions or statements of belief come under social influence.

Conformity bias, meanwhile, reflects the fact that completely separately from our rational judgments, we simply do not like to stick out from a pack. It makes us feel bad. The transcripts of the Asch experiment are telling. Even subjects who went against the grain and trusted their own judgment expressed discomfort with doing so. Here is how Asch described a subject who bucked the trend and chose the correct line: "His later answers came in a whispered voice, accompanied by a deprecating smile. At one point he grinned embarrassedly, and whispered explosively to his neighbor: 'I always disagree—darn it!'"[71]

Conformity bias can help explain what happened when Semmelweis showed that hand-washing could prevent puerperal fever. His peers—none of whom were washing their hands—ignored him because they all agreed it was absurd to suppose that gentlemen's hands could transmit disease. Bolstered by their fellows, they were unwilling to countenance evidence to the contrary—even though the evidence was powerful and immediate. Likewise, Trump supporters' claims that a relatively empty picture has more people in it may stem from their desire to agree with those they associate with.[72]

In the model developed by Bala and Goyal, we saw that social ties can have a remarkable influence on how communities of scientists come to believe things. But the variations we have discussed so far have been based on the assumption that what each individual cares about is the truth—or at least, trying to take the best action. The research on conformity bias suggests that we care about more than just the best action. At least in some settings, it seems we also care about agreeing with other people. In fact, in some cases we are prepared to deny our beliefs, or the evidence of our senses, to better fit in with those around us.

How might an effect like conformity bias influence communities of scientists?[73] Suppose we begin with the basic Bala-Goyal model that we have already described. The scientists in the model update their beliefs in light of the results of their own actions and those of others, just as before. But now suppose that when the scientists in our models choose how to act, they do so in part on the basis of what those around them do.[74] We might suppose that they derive some payoff from agreeing with others and that this influences their decisions about which action to take, but that they also update their beliefs about the world on the basis of what they and their neighbors observe. We can imagine different scenarios—in some cases, or for some scientists, conformity might be very important, so that it heavily influences their choices. In other cases, they care more about the benefits of the better action, or prescribing a better drug, and so pay less attention to what their colleagues are doing.

In the extreme case, we can consider what happens when the *only* thing that scientists care about is conforming their actions to those of others—or at least, when the payoff from conforming is much larger than that from performing the better action. Under these conditions, the models predict that groups of scientists are just as likely to end up at a bad consensus as a good one. A group investigating puerperal fever is just as likely to settle on hand-washing as not. After all, if they only care about matching each other, the feedback they get from the world makes no difference at all. In this extreme case, social connections have a severe dampening effect on scientists' ability to reach true beliefs.

Worse, once they find an action they all agree on, they will keep performing that action regardless of any new evidence. They will do this even if *all* the scientists come to believe something else is actually better, because no one is willing to buck the consensus. Those without peers to worry about, on the other hand, are unhampered by a desire to conform and are willing to try out a new, promising theory.[75]

Of course, the assumption that scientists care *only* about conforming is too strong. As the Asch experiment shows, people care about conformity but also about truth. What about models in which we combine the two elements? Even for these partially truth-seeking scientists, conformity makes groups of scientists worse at figuring out what is true.

First, the greater scientists' desire to conform, the more cases there are in which some of them hold correct beliefs but do not act on them. In other words, it is entirely possible that some of Semmelweis's peers believed that hand-washing worked but decided not to adopt it for fear of censure. In a network in which scientists share knowledge, this is especially bad. Each doctor who decided not to try hand-washing himself deprived all of his friends and colleagues of evidence about its efficacy. Conformity nips the spread of good new ideas in the bud.

Of course, conformity can also nip the spread of bad ideas in the bud, but we find that, on average, the greater their tendencies to conform, the more often a group of scientists will take the worse action. When they care only about performing the best action, they converge to the truth most of the time. In other words, they are pretty good at figuring out that, yes, hand-washing is better. But the more they conform, the closer we get to the case in which scientists end up at either theory completely randomly, because they do not care about the payoff differences between them. Pressures from their social realm swamp any pressures from the world.

Adding conformity to the model also creates the possibility of stable, persistent disagreement about which theory to adopt. Remember that in the Bala-Goyal base model, scientists always reach consensus, either correct or not. But now imagine a scenario in which scientists are clustered in small, tight-knit groups that are weakly connected to each other.

This is not such an unusual arrangement. Philosopher of science

Mike Schneider points out that in many cases, scientists are closely connected to those in their own country, or who are part of their racial or ethnic group. He shows that when scientists care about conformity, these kinds of groups can be a barrier to the spread of new ideas.[76]

Our results support this. When we have cliques of scientists, we see scenarios in which one group takes the worse action (no hand-washing) and the other takes the better action (hand-washing), but since the groups are weakly connected, conformity within each group keeps them from ever reaching a common consensus. In such a case, there are members of the non-hand-washers who know the truth—the ones who get information from the other group—but they never act on it and so never spread it to their compatriots.

Figure 6 shows a picture of this. One group takes action B, and the other takes action A. Some in the A group think B is better, including those connected to the B group, but their actions are fixed by their desire to conform with the group they are most connected to.

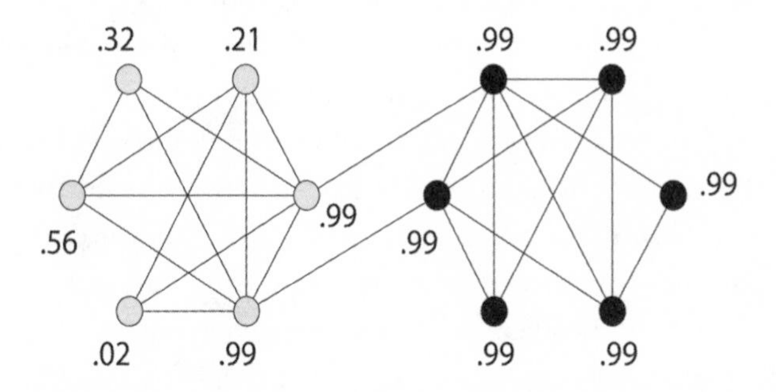

Figure 6. A cliquish arrangement of scientists with stable, opposing beliefs due to conformity. Light nodes represent individuals taking action A, and dark nodes, B. Within the A group, some individuals have accurate credences, that is, they believe B is better. Because they conform with the actions of their clique, this accurate belief is not transmitted to colleagues in that clique.

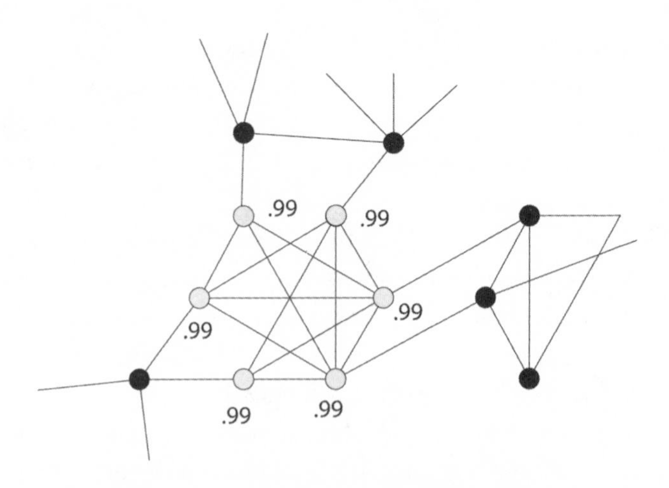

Figure 7. A network in which all agents have correct beliefs, but one clique takes the worse action because of conformity bias. Although individuals get information about the success of B from other colleagues, this is not enough for them to overcome their desire to conform with their group. Light nodes represent individuals taking action A, and dark nodes, B.

We can even find networks in which *everyone* holds the true belief—there is a real consensus in belief—but nonetheless, a large portion of scientists perform the worse action as a result of conformist tendencies.[77] Consider the partial network shown in figure 7. While the people in the central cluster all get good evidence about the benefits of hand-washing from those around them, because they are closely connected to each other, they are unwilling to change their practice. The light nodes all choose action A because of conformity, even though their connections to the rest of the network have led them to believe that B is better.

So, we see that the desire to conform can seriously affect the ability of scientists, or other people gathering knowledge, to arrive at good beliefs. Worse, as philosopher Aydin Mohseni and economist Cole Williams argue, knowing about conformity can also hurt scientists' ability to trust each other's statements.[78] If physicians say they are completely certain their hands do not carry cadaverous particles, it is hard to know whether they are convinced of this because of good evidence or because they are simply following the crowd.

Thus far, we have been thinking of "desire to conform" as the main variable in these models. But really, we should be looking at the trade-off between the desire to conform and the benefits of successful actions. In some situations the world pushes back so hard that it is nearly impossible to ignore, even when conformity is tempting. Suppose that, in our models, action B is much better than A. It pays off almost all the time, while A does so only rarely. In this sort of case, we find that agents in the models are more likely to disregard their desire to conform and instead make choices based on the best evidence available to them.

This can help explain why the Vegetable Lamb persisted for so long. There was almost no cost to believing the wrong thing. This means that *any* desire to conform could swamp the costs of holding a false belief. The wise medieval thinkers who waxed poetic about how delicious the Vegetable Lamb was derived social benefits from agreeing with their fellow literati. And the world never punished them for it.[79]

The Semmelweis case is different. There is no doubt that Semmelweis's hand-washing practice had dramatic real-world consequences—so it might seem surprising that physicians nonetheless conformed rather than try the promising new practice. But notice that the physicians themselves were not the ones at risk of death. Neither were their friends, relatives, or members of their social circles, as the patients in their clinics were generally poor. If the consequences of their choices were more personal, they might have ignored the reputational risks of admitting that their hands were unclean and listened to Semmelweis. Likewise, if we offered Trump supporters one thousand dollars to choose the photo with more people in it, more of them might get it right. (Relatedly, social psychologists have shown that when monetary incentives are offered to those who get the correct answer in the Asch test, conformity is less prevalent.)[80]

The difference between cases in which beliefs really matter and in which they are more abstract can help us understand some modern instances of false belief as well. When beliefs are *not* very important to action, they can come to take the role of a kind of social signal. They tell people what group you belong to—and help you get whatever benefits might accrue from membership in that group.

For example, an enormous body of evidence supports the idea that the biological species in our world today evolved via natural selection. This is the cornerstone of modern biology, and yet whether or not we accept evolution—irrespective of the evidence available—has essentially no practical consequences for most of us. On the other hand, espousing one view or the other can have significant social benefits, depending on whom we wish to conform with.

Likewise, there are a number of widely held, pseudoscientific beliefs about food and health that tend to have few negative consequences for those who hold them. Consider the beliefs that irradiated food is dangerous, fears of genetically modified foods, and beliefs that organic foods have special health benefits (beyond the lack of contamination from pesticides).[81] In each of these cases, there may be inconveniences and costs for consumers who avoid irradiated, genetically modified, or nonorganic foods, but these are relatively minor. At the same time, these eating practices can signal membership among new-age, elite, or left-wing social groups, and thus bring social benefits. These same communities sometimes promote wackier ideas—such as the recent fad of "grounding," based on claims that literally touching the ground provides health benefits as a result of electron transfer between the body and earth. Again, people are not going to be hurt by putting their feet on the ground (in fact walking barefoot is often relaxing and pleasant), and so such a belief should be easily stabilized by social influences.[82]

Thinking of false beliefs as social signals makes the most sense when we have cliquish networks like those in figure 6. When two cliques settle on two different beliefs, those beliefs come to signal group membership. A man who says he does not believe in evolution tells you something not just about his beliefs but about where he comes from and whom he identifies with.

Notice that the resulting arrangement can look an awful lot like polarization: there are two (or more) groups performing different actions (and perhaps with different beliefs), neither of which listens to the other. In both cases, there is no social influence between the groups. But perhaps surprisingly, the reasons are very different. In our polarization models, social influence fails because individuals stop trusting each other. In the conformity models, we see an outcome that, practically speaking, looks the same as polarization because everyone tries to conform with everyone else, but some people just do not interact very often. A glimpse back at figures 4 and 6 will make clear just how different these two outcomes really are.

The fact that polarization-like behavior can arise for very different reasons makes it especially hard to evaluate possible interventions. In the conformity case, disturbing people's social networks and connecting them with different groups should help rehabilitate those with false beliefs. But when people polarize because of mistrust, such an intervention would generally fail—and it might make polarization worse. In the real world, both effects seem to be at work, in which case interventions will need to be sensitive to both explanations for false belief.

We have seen in this chapter that the effects of social engagement on our beliefs and behaviors are myriad and complex. Our social networks are our best sources of new evidence and beliefs. But they also open us up to negative social effects. Friends and colleagues can help us learn about the health risks of eating fish and the best way to treat ulcers—but they leave us susceptible to Vegetable Lambs.

As we have argued, these social effects are often independent of, though sometimes exacerbated by, individual psychological tendencies. When we use the beliefs of others to ground our judgment of the evidence they share, we can learn to ignore those who might provide us with crucial information. When we try to conform to others in our social networks, we sometimes ignore our best judgment when making decisions, and, in doing so, halt the spread of true belief.

Things are about to get worse, however. So far, we have assumed that all of the scientists in our models share real results, and that they are all motivated by the goal of establishing truth. But the history of science—and politics—reveals that this is often a bad assumption. There are powerful forces in the world whose interests depend on public opinion and who manipulate the social mechanisms we have just described to further their own agendas.

THREE

The Evangelization of Peoples

In December 1952, *Reader's Digest* published an article titled "Cancer by the Carton," which presented the growing evidence of a link between cigarette smoking and lung cancer.[1] The article pulled no punches: it asserted that deaths from lung cancer had increased by a factor of 10 from 1920 to 1948 and that the risk of lung cancer in smokers older than forty-five increased in direct proportion to the number of cigarettes smoked. It quoted a medical researcher who speculated that lung cancer would soon become the most common form of human cancer—precisely because of the "enormous increase" in smoking rates per capita in the United States and elsewhere. Perhaps most important of all—at least from the perspective of the tobacco industry—the article called the increase in lung cancer "preventable" and suggested that the public needed to be warned of the dangers of smoking.

The article was a public relations doomsday scenario for the tobacco industry. At the time, *Reader's Digest* had a circulation of tens of millions of copies and was one of the most widely read publications in the world.[2] The dramatic headline, clear and concise prose, and unambiguous assessment of the scientific evidence made a greater impact than any public health campaign could have done. The article left no uncertainty: smokers were slowly killing themselves.

Soon more evidence came in. During the summer of 1953, a group of doctors at Sloan Kettering Memorial Hospital completed a study in which they painted mice with cigarette tar. The mice reliably developed malignant carcinomas.[3] Their paper provided a direct and visceral causal link between a known by-product of smoking and fatal cancer, where previous studies had shown only statistical relationships. It produced a media frenzy, with articles appearing in national and international newspapers and magazines. (*Time* magazine ran the story under the title "Beyond Any Doubt.")[4] That December, four more studies bolstering the case were presented at a research meeting in New York; one doctor told the *New York Times* that "the male population of the United States will be decimated by cancer of the lung in another fifty years if cigarette smoking increases as it has in the past."[5]

The bad press had immediate consequences. The day after it reported on the December research meeting, the *Times* ran an article contending that a massive sell-off in tobacco stocks could be traced to the recent coverage. The industry saw three consecutive quarters of decline in cigarette purchases, beginning shortly after the *Reader's Digest* article.[6] (This decline had followed nineteen consecutive quarters of record sales.) As National Institutes of Health statistician Harold Dorn would write in 1954, "Two years ago cancer of the lung was an unfamiliar and little discussed disease outside the pages of medical journals. Today it is a common topic of discussion, apparently throughout the entire world."[7]

The tobacco industry panicked. Recognizing an existential threat, the major US firms banded together to launch a public relations campaign to counteract the growing—correct—perception that their product was killing their customers. Over the two weeks following the market sell-off, tobacco executives held a series of meetings at New York's Plaza Hotel with John Hill, cofounder of the famed public relations outfit Hill & Knowlton, to develop a media strategy that could counter a steady march of hard facts and scientific results.

As Oreskes and Conway document in *Merchants of Doubt*, the key idea behind the revolutionary new strategy—which they call the "Tobacco Strategy"—was that the best way to fight science was with more science.[8]

Of course, smoking *does* cause lung cancer—and also cancers of the mouth and throat, heart disease, emphysema, and dozens of other serious illnesses. It would be impossible, using any legitimate scientific method, to generate a robust and convincing body of evidence demonstrating that smoking is safe. But that was not the goal. The goal was rather to create the appearance of uncertainty: to find, fund, and promote research that muddied the waters, made the existing evidence seems less definitive, and gave policy makers and tobacco users just enough cover to ignore the scientific consensus. As a tobacco company executive put it in an unsigned memo fifteen years later: "Doubt is our product since it is the best means of competing with the 'body of fact' that exists in the mind of the public."[9]

At the core of the new strategy was the Tobacco Industry Research Committee (TIRC), ostensibly formed to support and promote research on the health effects of tobacco. In fact it was a propaganda machine. One of its first actions was to produce, in January 1954, a document titled "A Frank Statement to Cigarette Smokers."[10] Signed by the presidents and chairmen of fourteen tobacco companies, the "Frank Statement" ran as an advertisement in four hundred newspapers across the United States. It responded to general allegations that tobacco was unsafe—and explicitly commented on the Sloan Kettering report that tobacco tar caused cancer in mice. The executives asserted that this widely reported study was "not regarded as conclusive in the field of cancer research" and that "there is no proof that cigarette smoking is one of the causes" of lung cancer. But they also claimed to "accept an interest in people's health as a basic responsibility, paramount to every other consideration" in their business. The new committee would provide "aid and assistance to the research efforts into all phases of tobacco use and health."

The TIRC did support research into the health effects of tobacco, but its activities were highly misleading. Its main goal was to promote scientific research that contradicted the growing consensus that smoking kills.[11] The TIRC sought out and publicized the research of scientists whose work was likely to be useful to them—for instance, those studying the links between lung cancer and other environmental factors, such as asbestos.[12] It produced pamphlets such as "Smoking and Health," which in 1957 was distributed to hundreds of thousands of doctors and dentists and which described a very biased sample of the available research on smoking. It consistently pointed to its own research as evidence of an ongoing controversy over the health effects of tobacco and used that putative controversy to demand equal time and attention for the industry's views in media coverage.

This strategy meant that even as the scientific community reached consensus on the relationship between cigarettes and cancer—including, as early as 1953, the tobacco industry's *own* scientists—public opinion remained torn.[13] After significant drops in 1953 and into 1954, cigarette sales began rising again and did so steadily for more than two decades—until long after the science on the health risks of tobacco was completely settled.[14]

In other words, the Tobacco Strategy worked.

•

The term "propaganda" originated in the early seventeenth century, when Pope Gregory XV established the Sacra Congregatio de Propaganda Fide—the Sacred Congregation for the Propagation of the Faith. The Congregation was charged with spreading Roman Catholicism through missionary work across the world and, closer to home, in heavily Protestant regions of Europe. (Today the same body is called the Congregation for the Evangelization of Peoples.) Politics and religion were deeply intertwined in seventeenth-century Europe, with major alliances and even empires structured around theological divides between Catholics and Protestants.[15] The Congregation's activities within Europe were more than religious evangelization: they amounted to political subversion, promoting the interests of France, Spain, and the southern states of the Holy Roman Empire in the Protestant strongholds of northern Europe and Great Britain.

It was this political aspect of the Catholic Church's activities that led to the current meaning of propaganda as the systematic, often biased, spread of information for political ends. This was the sense in which Marx and Engels used the term in the *Communist Manifesto*, when they said that the founders of socialism and communism sought to create class consciousness "by our propaganda."[16] Joseph Goebbels's title as "propaganda minister" under the German Third Reich also invokes this meaning. The Cold War battles between the United States and the Soviet Union over "hearts and minds" are aptly described as propaganda wars.

Many of the methods of modern propaganda were developed by the United States during World War I. From April 1917 until August 1919, the Committee on Public Information (CPI) conducted a systematic campaign to sell US participation in the war to the American public.[17] The CPI produced films, posters, and printed publications, and it had offices in ten countries, including the United

States. In some cases it fed newspapers outright lies about American activities in Europe—and occasionally got caught, leading the *New York Times* to run an editorial calling it the Committee on Public Misinformation. One member of the group later described its activities as "psychological warfare."[18]

After the war, the weapons of psychological warfare were turned on US and Western European consumers. In a series of books in the 1920s, including *Crystallizing Public Opinion* (1923) and *Propaganda* (1928), CPI veteran Edward Bernays synthesized results from the social sciences and psychology to develop a general theory of mass manipulation of public opinion—for political purposes, but also for commerce.

Bernays's postwar work scarcely distinguished between the political and commercial. One of his most famous campaigns was to rebrand cigarettes as "torches of freedom," a symbol of women's liberation, with the goal of breaking down social taboos against women's smoking and thus doubling the market for tobacco products. In 1929, under contract with the American Tobacco Company, makers of Lucky Strike cigarettes, he paid women to smoke while marching in the Easter Sunday Parade in New York.

The idea that industry—including tobacco, sugar, corn, healthcare, energy, pest control, firearms, and many others—is engaged in propaganda, far beyond advertising and including influence and information campaigns addressed at manipulating scientific research, legislation, political discourse, and public understanding, can be startling and deeply troubling. Yet the consequences of these activities are all around us.

Did (or do) you believe that fat is unhealthy—and the main contributor to obesity and heart disease? The sugar industry invested heavily in supporting and promoting research on the health risks of fat, to deflect attention from the greater risks of sugar.[19] Who is behind the long-term resistance to legalizing marijuana for recreational use? Many interests are involved, but alcohol trade groups have taken a particularly strong and effective stand.[20] There are many examples of such industry-sponsored beliefs, from the notion that opioids prescribed for acute pain are not addictive to the idea that gun owners are safer than people who do not own guns.[21]

Bernays himself took a rosy view of the role that propaganda, understood to include commercial and industrial information campaigns, could play in a democratic society. In his eyes it was a tool for beneficial social change: a way of promoting a more free, equal, and just society. He particularly focused on how propaganda could aid the causes of racial and gender equality and education reform. Its usefulness for lining the pockets of Bernays and his clients was simply another point in its favor. After all, he was writing in the United States at the peak of the Roaring Twenties; he had no reason to shy away from capitalism. Propaganda was the key to a successful democracy.

Today it is hard not to read Bernays's books through the lens of their own recommendations, as the work of a public relations spokesman for the public relations industry. And despite his reassurances, a darker side lurks in his pages. He writes, for instance, that "those who manipulate this unseen mechanism of society constitute an invisible government which is the true ruling power of our country. We are governed, our minds molded, our tastes formed, our ideas suggested, largely by men we have never heard of."[22] This might sound like the ramblings of a conspiracy theorist, but in fact it is far more nefarious: it is an invitation to the conspiracy, drafted by one of its founding fathers, and targeted to would-be titans of industry who would like to have a seat on his shadow council of thought leaders.

Perhaps Bernays overstated his case, but his ideas have deeply troubling consequences. If he is right, then the very idea of a democratic society is a chimera: the will of the people is something to be shaped by hidden powers, making representative government meaningless. Our only hope is to identify the tools by which our beliefs, opinions, and preferences are shaped, and look for ways to re-exert control—but the success of the Tobacco Strategy shows just how difficult this will be.

The Tobacco Strategy was wildly successful in slowing regulation and obscuring the health risks of smoking. Despite strong evidence of the link between smoking and cancer by the early 1950s, the Surgeon General did not issue a statement linking smoking to health risks until 1964—a decade after the TIRC was formed.[23] The following year, Congress passed a bill requiring a health warning on tobacco products. But it was not until 1970 that cigarette advertising was curtailed at the federal level, and not until 1992 that the sale of tobacco products to minors was prohibited.[24]

All of this shows that the industry had clear goals, it adopted a well-thought-out strategy to accomplish them, and the goals were ultimately reached. What is much harder to establish by looking at the history alone is the degree to which the Tobacco Strategy contributed to those goals. Did industry propaganda make a difference? If so, which aspects of its strategy were most effective?

There were many reasons why individuals and policy makers might prefer to delay regulation and disregard the evidence about links between cancer and smoking. Was anyone ever as cool as Humphrey Bogart with a cigarette hanging from his lips? Or Audrey Hepburn with an arm-length cigarette holder? Smoking was culturally ubiquitous during the 1950s and 1960s, and it was difficult to imagine changing this aspect of American society by government order. Worse, many would-be regulators of the tobacco industry were smokers themselves. Clear conflicts of interest arise, independently of any industry intervention, when the users of an addictive product attempt to regulate the industry that produces it. And in addition to funding research, the tobacco industry poured millions of dollars into funding political campaigns and lobbying efforts.

Subtle sociological factors can also influence smoking habits in ways that are largely independent of the Tobacco Strategy. In a remarkable 2008 study, Harvard public health expert Nicholas Christakis and UC San Diego political scientist James Fowler looked at a social network of several thousand subjects to see how social ties influenced their smoking behavior.[25] They found that smokers often cluster socially: those with smoking friends were more likely to be smokers and vice versa. They also found that individuals who stopped smoking had a big effect on their friends, on friends of friends, and even on friends of friends of friends. Clusters of individuals tended to stop smoking together. Of course, the converse is that groups who keep smoking tend to do so together as well.[26] When the cancer risks of cigarettes were first becoming clear, roughly 45 percent of US adults were smokers. Who wants to be the first to stop?[27]

To explore in more detail how propagandists can manipulate public belief, we turn once again to the models we looked at in the last chapter. We can adapt them to ask: Should we expect the Tobacco Strategy and similar propaganda efforts to make a significant difference in public opinion? Which features of the Tobacco Strategy are most effective, and why do they work? How can propaganda combat an overwhelming body of scientific work? Working with Justin Bruner, a philosopher and political scientist at the Australian National University, and building on his work with philosopher of science Bennett Holman (which we discuss later), we have developed a model that addresses these questions.[28]

We begin with the basic Bala-Goyal model we described in the last chapter. This, remember, involves a group of scientists who communicate with those in their social network. They are all trying to figure out whether one of two actions—A or B—will yield better results on average. And while they know exactly how often A leads to good outcomes, they are unsure about whether B is better or worse than A. The scientists who think A is better perform that action, while those who lean toward B test it out. They use Bayes' rule to update their beliefs on the basis of the experiments they and their colleagues perform. As we saw, the most common outcome in this basic model is that the scientific community converges on the better theory, but the desire to conform and the mistrust of those with different beliefs can disrupt this optimistic picture.

In the last chapter, we used this model and variations on it to understand social effects in scientific communities. But we can vary the model to examine how ideas and evidence can flow from a scientific community to a community of nonscientists, such as policy makers or the public, and how tobacco strategists can interfere with this process.

We do this by adding a new group of agents to the model, whom we call policy makers. Like scientists, policy makers have beliefs, and they use Bayes' rule to update them in light of the evidence they see. But unlike scientists, they do not produce evidence themselves and so must depend on the scientific network to learn about the world. Some policy makers might listen to just one scientist, others to all of them or to some number in between.

Figure 8 shows this addition to the model. On the right we have our community of scientists as before, this time arranged in a cycle. As before, some of them favor theory A (the light nodes) and others B (the dark ones). On the left, we add policy makers (squares instead of circles), each with their own belief about whether theory B is better than A. The dotted lines indicate that while they have connections with scientists, these are one-sided. This figure shows one policy maker who listens to a single scientist, one who listens to two, and one who listens to three.

With just this modification to the framework, we find that policy

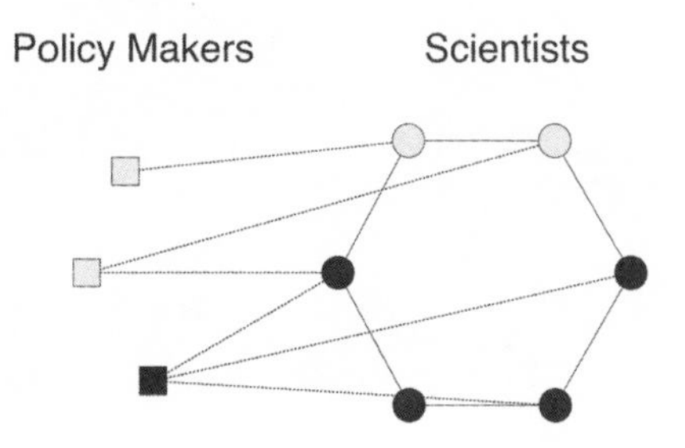

Figure 8. An epistemic network with policy makers and scientists. Although both groups have beliefs about whether action A (light nodes) or action B (dark nodes) is better, only scientists actually test these actions. Policy makers observe results from some set of scientists and update their beliefs. Dotted lines between scientists and policy makers reflect this one-sided relationship.

makers' beliefs generally track scientific consensus. Where scientists come to favor action B, policy makers do as well. This occurs even if the policy makers are initially skeptical, in the sense that they start with credences that favor A (less than .5). When policy makers are connected to only a small number of scientists, they may approach the true belief more slowly, but they always get there eventually (as long as the scientists do).

Now consider what happens when we add a propagandist to the mix. The propagandist is another agent who, like the scientists, can share results with the policy makers. But unlike the scientists, this agent is not interested in identifying the better of two actions. This agent aims only to persuade the policy makers that action A is preferable—even though, in fact, action B is. Figure 9 shows the model with this agent. Propagandists do not update their beliefs, and they communicate with every policy maker.

The Tobacco Strategy was many faceted, but there are a handful of specific ways in which the TIRC and similar groups used science to

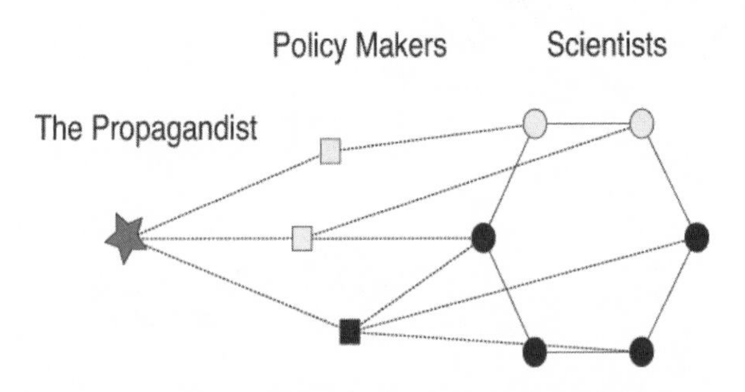

Figure 9. An epistemic network with scientists, policy makers, and a propagandist. The propagandist does not hold beliefs of their own. Instead, their goal is to communicate only misleading results to all the policy makers. Light nodes represent individuals who prefer action A, and dark nodes, B.

fight science.[29] The first is a tactic that we call "biased production." This strategy, which may seem obvious, involves directly funding, and in some cases performing, industry-sponsored research. If industrial forces control the production of research, they can select what gets published and what gets discarded or ignored. The result is a stream of results that are biased in the industry's favor.

The tobacco companies invested heavily in such research following the establishment of the TIRC. By 1986, according to their own estimates, they had spent more than $130 million on sponsored research, resulting in twenty-six hundred published articles.[30] Awards were granted to researchers and projects that the tobacco industry expected to produce results beneficial to them. This research was then shared, along with selected independent research, in industry newsletters and pamphlets; in press releases sent to journalists, politicians, and medical professionals; and even in testimony to Congress.

Funding research benefited the industry in several ways. It provided concrete (if misleading) support for tobacco executives' claims that they cared about smokers' health. It gave the industry access to doctors who could then appear in legal proceedings or as industry-friendly experts for journalists to consult. And it produced data that could be used to fight regulatory efforts.

It is hard to know what sorts of pressure the tobacco industry placed on the researchers it funded. Certainly the promise of future funding was an incentive for those researchers to generate work that would please the TIRC. But there is strong evidence that the tobacco industry itself produced research showing a strong link between smoking and lung cancer that it did not publish. Indeed, as we noted, the industry's own scientists appear to have been convinced that smoking causes cancer as early as the 1950s—and yet the results of those studies remained hidden for decades, until they were revealed through legal action in the 1990s. In other words, industry scientists were not only producing studies showing that smoking was safe, but when their studies linked tobacco and cancer, they buried them.

Let us add this sort of biased production to our model. In each round, we suppose the propagandist always performs action B but then shares only those outcomes that happen to suggest action A is better. Suppose that in each study, the propagandist takes action B ten times. Whenever this action is successful four times or fewer, they share the results. Otherwise not. This makes it look like action B is, on average, worse than A (which tends to work five times out of ten). The policy makers then update their beliefs on this evidence using Bayes' rule. (The policy makers also update their beliefs on any results shared by the scientists they are connected to, just as before.)[31]

Figure 10 gives an example of what this might look like. In (a) we see that the policy makers have different credences (scientist credences are omitted from the figure for simplicity's sake). In (b) both the propagandist and the scientists test their beliefs. Scientists "flip the coin" ten times each. The propagandist, in this example,

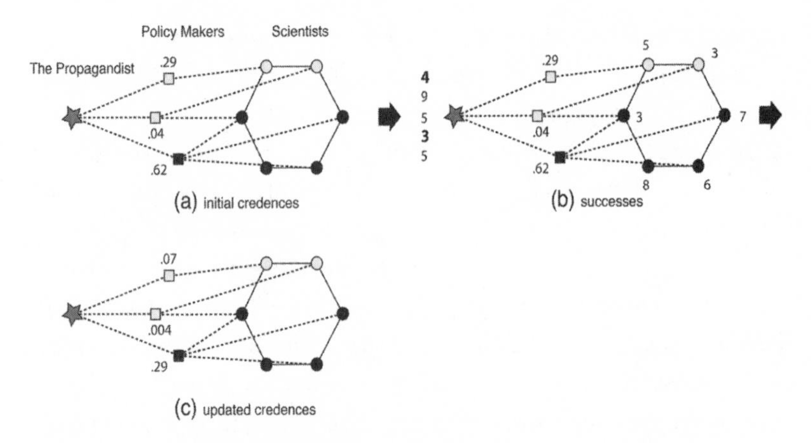

Figure 10. An example of policy-maker belief updating in a model in which a propagandist engages in biased production. In (a) we see the initial credences of the policy makers. In (b) the scientists test their beliefs, and the propagandist tests theory B. The propagandist chooses to share only those trials (bolded) that spuriously support A as the better theory. In (c) we see how policy makers update their credences in light of this evidence. Light nodes represent individuals who prefer action A, and dark nodes, B.

has enough funding to run five studies, each with ten test subjects, and so sees five results. Then, the scientists share their results, and the propagandist shares just the two bolded results, that is, the ones in which B was not very successful. In (c) the policy makers have updated their beliefs.

We find that this strategy can drastically influence policy makers' beliefs. Often, in fact, as the community of scientists reaches consensus on the correct action, the policy makers approach certainty that the *wrong* action is better. Their credence goes in precisely the wrong direction. Worse, this behavior is often stable, in the sense that no matter how much evidence the scientific community produces, as long as the propagandist remains active, the policy makers will never be convinced of the truth.

Notice that in this model, the propagandist does not fabricate any data. They are performing real science, at least in the sense that they actually perform the experiments they report, and they do so using the same standards and methods as the scientists. They just publish the results selectively.

Even if it is not explicit fraud, this sort of selective publication certainly seems fishy—as it should, since it is designed to mislead. But it is important to emphasize that selective publication is common in science even without industrial interference. Experiments that do not yield exciting results often go unpublished, or are relegated to minor journals where they are rarely read.[32] Results that are ambiguous or unclear get left out of papers altogether. The upshot is that what gets published is never a perfect reflection of the experiments that were done. (This practice is sometimes referred to as "publication bias" or the "file drawer effect," and it causes its own problems for scientific understanding.)[33] This observation is not meant to excuse the motivated cherry-picking at work in the biased production strategy. Rather, it is to emphasize that, for better or worse, it is continuous with ordinary scientific practice.[34]

One way to think about what is happening in these models is that there is a tug-of-war between scientists and the propagandist for the hearts and minds of policy makers. Over time, the scientists' evidence will tend to recommend the true belief: more studies will support action B because it yields better results on average. So the results the scientists share will, on average, lead the policy makers to the true belief.

On the other hand, since the propagandist shares only those results that support the worse theory, their influence will always push policy makers' beliefs the other way. This effect always slows the policy makers' march toward truth, and if the propagandist manages to pull hard enough, they can reverse the direction in which the policy makers' credences move. Which side can pull harder depends on the details of the scientific community and the problem that scientists tackle.

For instance, it is perhaps unsurprising that industrial propagandists are less effective when policy makers are otherwise well-informed. The more scientists the policy makers are connected to, the greater the chance that they get enough evidence to lead them to the true theory. If we imagine a community of doctors who scour the medical literature for the dangers of tobacco smoke, we might expect them to be relatively unmoved by the TIRC's work. On the other hand, when policy makers have few independent connections to the scientific community, they are highly vulnerable to outside influence.

Likewise, the propagandist does better if they have more funding. More funding means that they can run more experiments, which are likely to generate more spurious results that the propagandist can then report.

Perhaps less obvious is that given some fixed amount of funding, how the propagandist chooses to allocate the funds to individual studies can affect their success. Suppose the propagandist has enough money to gather sixty data points—say, to test sixty smokers for cancer. They might allocate these resources to running one study with sixty subjects. Or they might run six studies with ten subjects each, or thirty studies, each with only two data points. Surprisingly, the propagandist will be most effective if they run and publicize the most studies with as few data points as possible.

Why would this be the case? Imagine flipping a coin that comes up heads 70 percent of the time, and you want to figure out whether it is weighted toward heads or toward tails. (This, of course, is analogous to the problem faced by scientists in our models.) If you flip this coin sixty times, the chances are very high that there will be more heads overall. Your study is quite likely to point you in the right direction. But if you flip the coin just once, there is a 30 per-

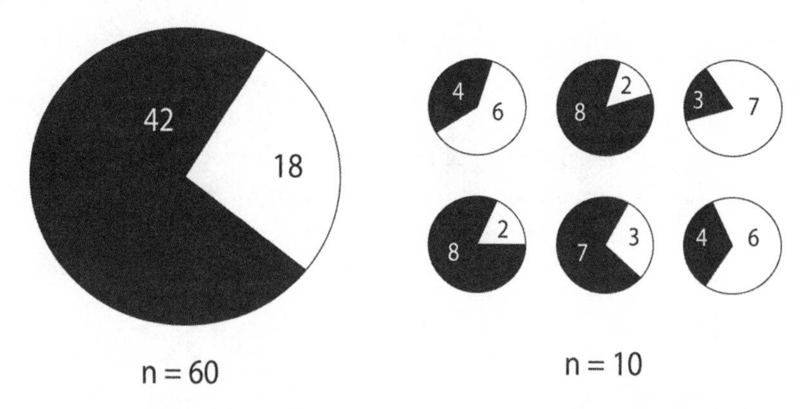

Figure 11. Breaking one large study into many smaller ones can provide fodder for propagandists. On the left we see a trial with sixty data points, which reflects the underlying superiority of action B (dark). On the right, we see the same data points separated into six trials. Three of these spuriously support action A (light). A propagandist can share only these studies and mislead policy makers, which would not be possible with the larger study.

cent chance that your study will mislead you into thinking the coin is weighted toward tails. In other words, the more data points you gather, the higher the chances that they will reflect the true effect.

Figure 11 shows an example of this. It represents possible outcomes for one study with sixty subjects and for six studies with ten subjects each. In both cases, we assume that the samples exactly represent the real distribution of results for action B, meaning that the action worked 70 percent of the time. In other words, the results are the same, but they are broken up differently. While the sixty-subject study clearly points toward the efficacy of B, three of the smaller studies point toward A. The propagandist can share just these three and make it look as if the total data collected involved nineteen failures of B and only eleven successes.

Likewise, in an extreme case the propagandist could use their money to run sixty studies, each with only one data point. They would then have the option to report all and only the studies in which action B failed—without indicating how many times they flipped the coin and got the *other* result. An observer applying Bayes' rule with just this biased sample of the data would come away thinking that action A is much better, even though it is actually worse. In other words, the less data they have, the better the chances that each result is spurious—leading to more results that the propagandist can share.

Biased production is in many ways a crude tool—less crude, perhaps, than outright fraud, but not by much. It is also risky. A strategy like this, if exposed, makes it look as if the propagandist has something to hide. (And indeed, they do—all the studies tucked away in their file drawers.) But it turns out that the propagandist can use more subtle tools that are both cheaper and, all things considered, more effective. One is what we call "selective sharing." Selective sharing involves searching for and promoting research that is conducted by independent scientists, with no direct intervention by the propagandist, that happens to support the propagandist's interests.

As Oreskes and Conway show, selective sharing was a crucial component of the Tobacco Strategy. During the 1950s, when a growing number of studies had begun to link smoking with lung cancer, a group called the Tobacco Institute published a regular newsletter called *Tobacco and Health* that presented research suggesting there was no link.[35] This newsletter often reported independent results, but in a misleadingly selective way—with the express purpose of undermining other widely discussed results in the scientific literature.

For instance, in response to the Sloan Kettering study showing that cigarette tar produced skin cancer in mice, *Tobacco and Health* pointed to later studies by the same group that yielded lower cancer incidences, implying that the first study was flawed but not giving a complete account of the available data. The newsletter ran headlines such as "Five Tobacco-Animal Studies Report No Cancers Induced" without mentioning how many studies *did* report induced cancers. This strategy makes use of a fundamental public misunderstanding of how science works. Many people think of individual scientific studies as providing proof, or confirmation, of a hypothesis. But the probabilistic nature of evidence means that real science is far from this ideal. Any one study can go wrong, a fact Big Tobacco used to its advantage.

Tobacco and Health also reported on links between lung cancer and other substances, such as asbestos, automobile exhaust, coal smoke, and even early marriage, implying that the recent decades' rise in lung cancer rates could have been caused by any or all of these other factors.

A closely related strategy involved extracting and publishing quotations from research papers and books that, on their face, seemed to express uncertainty or caution about the results. Scientists are sometimes modest about the significance of their studies, even when their research demonstrates strong links. For instance, Richard Doll, a British epidemiologist who conducted one of the earliest studies establishing that smoking causes cancer, was quoted in *Tobacco and Health* as writing, "Experiments in which animals were exposed to the tar or smoke from tobacco have uniformly failed to produce any pulmonary tumors comparable to the bronchial carcinoma of man."[36] The newsletter did not mention that these experiments *had* shown that animals exposed to tar and smoke would get carcinomas elsewhere. Doll was in fact drawing precise distinctions, but industry painted him as expressing uncertainty.[37]

In our model, the propagandist implements selective sharing by searching through the results produced by the scientific community and then passing along all and only those that happen to support their agenda. In many ways this ends up looking like biased production, in that the propagandist is sharing only favorable results, but with a big difference. The *only* results that get shared in this model are produced by independent researchers. The propagandist does not do science. They just take advantage of the fact that the data produced by scientists have a statistical distribution, and there will generally be some results suggesting that the wrong action is better.

Figure 12 shows an example of what happens to policy-maker beliefs under selective sharing by the propagandist.[38] (We reduce the number of scientists and policy makers in this figure to keep things legible.) Notice here that the propagandist is now observing the beliefs of all the scientists as well as communicating with the policy makers. In (a) we see their initial credences. In (b) the scientists test only their preferred actions, with three of them trying B. Two scientists happen to observe only four successes in this case, which each make it look like B is worse than A. The propagandist shares only these two results. In (c) we can see that, as a result, the policy makers now have less accurate beliefs.

In this strategy, the propagandist does absolutely nothing to interfere with the scientific process. They do not buy off scientists or fund their own research. They simply take real studies, produced independently, that by chance suggest the wrong answer. And they forward these and only these studies to policy makers.

It turns out that selective sharing can be extremely effective. As with biased production, we find that in a large range of cases, a propagandist using only selective sharing can lead policy makers to converge to the false belief even as the scientific community converges to the true one. This may occur even though the policy makers are *also* updating their beliefs in light of evidence shared by the scientists themselves.

The basic mechanism behind selective sharing is similar to that behind biased production: there is a tug-of-war. Results shared by

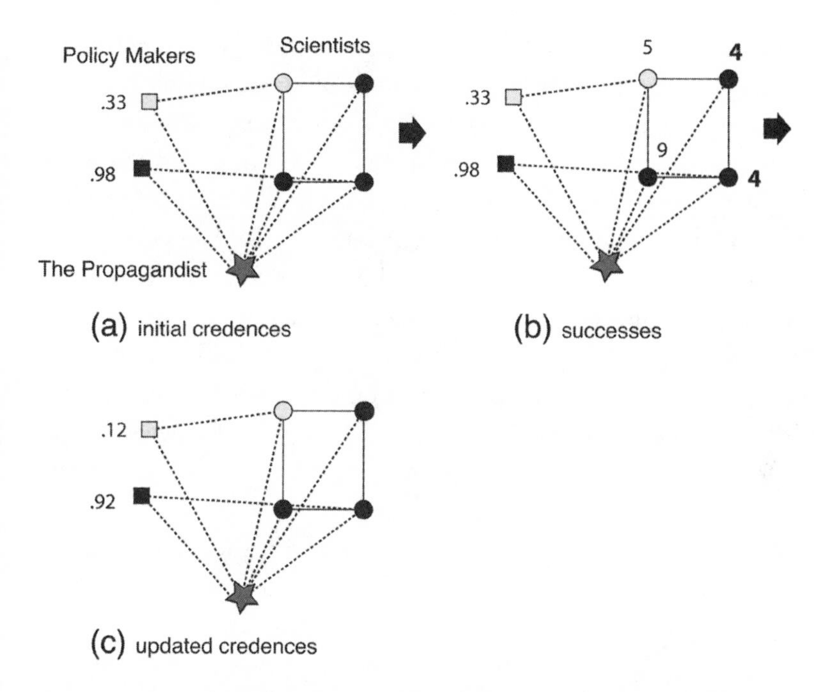

Figure 12. An example of policy-maker belief updating under selective sharing by a propagandist. In (a) we see the initial credences of the policy makers. In (b) scientists test their preferred actions. (Light nodes represent individuals taking action A, dark ones, B.) Some of these tests happen to spuriously support theory A, and the propagandist chooses only these (bolded) to share with policy makers. In (c) we see that policy makers have updated their beliefs on the basis of both the evidence directly from scientists and the spurious results from the propagandist.

scientists tend to pull in the direction of the true belief, and results shared by the propagandist pull in the other direction. The difference is that how hard the propagandist pulls no longer depends on how much money they can devote to running their own studies, but only on the rate at which spurious results appear in the scientific community.

For this reason, the effectiveness of selective sharing depends on the details of the problem in question. If scientists are gathering data on something where the evidence is equivocal—say, a disease in which patients' symptoms vary widely—there will tend to be more results suggesting that the wrong action is better. And the more misleading studies are available, the more material the propagandist has to publicize. If every person who smoked had gotten lung cancer, the Tobacco Strategy would have gone nowhere. But because the connections between lung cancer and smoking are murkier, tobacco companies had something to work with. (Something similar is true in the biased production case: generally, to produce the same number of spurious results, the propagandist needs to perform more studies—that is, spend more money—as evidence gets less equivocal.)

The practices of the scientific community can also influence how effective selective sharing will be—even if industry in no way interferes with the scientific process. The propagandist does especially well when scientists produce many studies, each with relatively little data. How much data is needed to publish a paper varies dramatically from field to field. Some fields, such as particle physics, demand extremely high thresholds of data quantity and quality for publication of experimental results, while other fields, such as neuroscience and psychology, have been criticized for having lower standards.[39]

Why would sparse data gathering help the propagandist? The answer is closely connected to why studies with fewer participants are better for the propagandist in the biased production strategy. If every scientist "flips their coin" one hundred times for each study, the propagandist will have very few studies to publicize, compared with a situation in which each scientist flips their coin, say, five times. The lower the scientific community's standards, the easier it is for the propagandist in the tug-of-war for public opinion.[40]

Of course, the more data you demand, the more expensive each study becomes, and this sometimes makes doing studies with more data prohibitive. But consider the difference between a case in which one scientist flips the coin one hundred times, and the propagandist can choose whether to share the result or not; and a case in which twenty scientists flip the coin five times each, and the propagandist can pick and choose among the studies. In the latter case, the propagandist's odds of finding something to share are much higher—even though the entire group of scientists in both cases performed the same total number of flips. Mathematically, this is the same as what we saw in figure 11, but for an independent community.

This observation leads to a surprising lesson for how we should fund and report science. You might think it is generally better to have more scientists work on a problem, as this will generate more ideas and greater independence. But under real-world circumstances, where a funding agency has a fixed pot of money to devote to a scientific field, funding more scientists is not always best. Our models suggest that it is better to give large pots of money to a few groups, which can use the money to run studies with more data, than to give small pots of money to many people who can each gather only a few data points. The latter distribution is much more likely to generate spurious results for the propagandist.

Of course, this reasoning can go too far. Funding just one scientist raises the risk of choosing someone with an ultimately incorrect hypothesis. Furthermore, there are many benefits to a more democratically structured scientific community, including the presence of a diversity of opinions, beliefs, and methodologies. The point here is that simply adding more scientists to a problem can have downsides when a propagandist is at work. Perhaps the best option is to fund many scientists, but to publish their work only in aggregation, along with an assessment of the total body of evidence.

Most disciplines recognize the importance of studies in which more data are gathered. (All else being equal, studies with more data are said to have higher statistical power, which is widely recognized as essential to rigorous science.)[41] Despite this, low-powered studies seem to be a continuing problem.[42] The prevalence of such studies is related to the so-called replication crisis facing the behavioral and medical sciences. In a widely reported 2010 study, a group of psychologists were able to reproduce only thirty-six of one hundred published results from their field.[43] In a 2016 poll run by the journal *Nature*, 70 percent of scientists across disciplines said they had failed to reproduce another scientist's result (and 50 percent said they had failed to reproduce a result of their own).[44] Since replicability is supposed to be a hallmark of science, these failures to replicate are alarming.

Part of the problem is that papers showing a novel effect are easier to publish than those showing no effect. Thus there are strong personal incentives to adopt standards that sometimes lead to spurious, but surprising, results. Worse, since, as discussed, studies that show no effect often never get published at all, it can be difficult to recognize which published results are spurious. Another part of the problem is that many journals accept underpowered studies in which spurious results are more likely to arise. This is why an interdisciplinary research team has recently advocated tighter minimal standards of publishability.[45]

Given this background, the possibility that studies with less data are fodder for the Tobacco Strategy is worrying. It goes without saying that we want our scientific communities to follow the practices most likely to generate accurate conclusions. That demanding experimental studies with high statistical power makes life difficult for propagandists only adds to the argument for more rigorous standards.

The success of selective sharing is striking because, given that it is such a minimal intervention into the scientific process, arguably it is not an intervention at all. In fact, in some ways it is even *more* effective than biased production, for two reasons. One is that it is much cheaper: the industry does not need to fund science, just publicize it. It is also less risky, because propagandists who share selectively do not hide or suppress any results. Furthermore, the rate at which the community of scientists will produce spurious results will tend to scale with the size of the community, which means that as more scientists work on a problem, the more spurious results they will produce, even if they generally produce more evidence for the true belief. Biased production, on the other hand, quickly becomes prohibitively expensive as more scientists join the fray.

Given the advantages to selective sharing, why does industry bother funding researchers at all? It turns out that funding science can have more subtle effects that shift the balance within the scientific community, and ultimately make selective sharing more effective as well.

In the summer of 2003, the American Medical Association (AMA) was scheduled to vote on a resolution, drafted by Jane Hightower, that called for national action on methylmercury levels in fish. The resolution would have demanded large-scale tests of mercury levels and a public relations campaign to communicate the results to the public. But the vote never took place.

As Hightower reports in her book *Diagnosis Mercury*, the day the resolution was set to be heard, the California delegate responsible for presenting it received word of a "new directive" stating that mercury in fish was not harmful to people. This new directive never actually materialized—but somehow the mere rumor of new evidence was enough to derail the hearing. Rather than vote on the resolution, the committee responsible passed it along to the AMA's

Council for Scientific Affairs for further investigation, meaning at least a year's wait before the resolution could be brought to the floor again.[46]

A year later, at the AMA's 2004 meeting, the council reported back. After extensive study, it concurred with the original resolution, recommending that fish be tested for methylmercury and the results publicly reported, and then went further—resolving that the FDA require the results of this testing to be posted wherever fish is sold.

But how did a rumor derail the presentation of the resolution in the first place? Where had the rumor come from?

This mystery is apparently unsolved, but in trying to understand what had happened, Hightower began to dig deeper into the handful of scientific results purporting to show that methylmercury in fish was not harmful after all. One research group in particular stood out. Based at the University of Rochester, this group had run a longitudinal study on a population with high fish consumption in the African nation of Seychelles. The researchers were investigating the possible effects of methylmercury on child development by comparing maternal mercury levels with child development markers.[47] They had published several papers showing no effect of methylmercury on the children involved—even as another large longitudinal study in the Faroe Islands reported the opposite result.[48]

Not long after the AMA meeting at which the resolution was originally scheduled to be discussed, a member of the Rochester group named Philip Davidson gave a presentation on the group's research. A friend of Hightower's faxed her a copy of the presentation, noting that the acknowledgments thanked the Electric Power Research Institute (EPRI)—a lobbying entity for the power industry, including the coal power industry responsible for the methylmercury in fish.[49]

Hightower discovered that the EPRI had given a $486,000 grant to a collaborative research project on methylmercury that included the Seychelles study. The same project had received $10,000 from the National Tuna Foundation and $5,000 from the National Fisheries Institute.[50] And while Davidson had thanked the EPRI in his presentation, he and his collaborators had not mentioned this funding in several of their published papers on children and methylmercury.

Hightower writes that she turned this information over to the Natural Resources Defense Council (NRDC) in Washington, D.C. After further investigation, the NRDC wrote to several journals that had published the Seychelles research, noting the authors' failure to reveal potential conflicts of interest. Gary Myers, another member of the Rochester group, drafted a response arguing that although the EPRI had funded the group, the papers in question were supported by other sources. The EPRI and fisheries interests, he wrote, "played no role in this study nor did they have any influence upon data collection, interpretation, analysis or writing of the manuscript."[51]

One might be skeptical that a significant grant from the coal industry would not influence research into whether the by-products of coal power plants affect child development. But we have no reason to think that the Rochester group did not act in good faith. So let us assume that the researchers were, at every stage, able to perform their work exactly as they would have without any industrial influence. Might industry funding have still had an effect? Philosophers of science Bennett Holman and Justin Bruner have recently argued that the answer is "yes": the mere fact that certain scientists received industry funding can dramatically corrupt the scientific process.

Holman and Bruner contend that industry can influence science without biasing scientists themselves by engaging in what they call "industrial selection." Imagine a community of scientists working on a single problem where they are trying to decide which of two actions is preferable. (They work in the same Bala-Goyal modeling framework we have already described; once again, assume that action A is worse than action B.) One might expect these scientists, at least initially, to hold different beliefs and hypotheses, and even to perform different sorts of tests on the world. Suppose further that some scientists use methods and hold background beliefs that are more likely to erroneously favor action A over action B.

To study this possibility, Holman and Bruner use a model in which each scientist "flips a coin" with a different level of bias. Most coins correctly favor action B, but some happen to favor A. The idea is that one could adopt methodologies in science that are not particularly well-tuned to the world, even if, on balance, most methods are. (Which methods are best is itself a subtle question in science.) In addition, Holman and Bruner assume that different practices mean that some scientists will be more productive than others. Over time, some scientists leave the network and are replaced—a regular occurrence when scientists retire or move on to other things. And it is more likely, in their models, that the replacement scientists will imitate the methods of the most productive scientists already in the network.

This sort of replacement dynamic is not a feature of the other models we have discussed. But the extra complication makes the models in some ways more realistic. The community of scientists in this model is a bit like a biological population undergoing natural selection: scientists who are more "fit" (in this case, producing more results) are also better at reproducing—that is, replicating themselves in the population by training successful students and influencing early-career researchers.[52]

Holman and Bruner also add a propagandist to the model. This time, however, the propagandist can do only one thing: dole out research money. The propagandist finds the scientist whose methods are most favorable for the theory they wish to promote and gives that scientist enough money to increase his or her productivity. This does two things. It floods the scientific community with results favorable to action A, changing the minds of many other scientists. And it also makes it more likely that new labs use the methods that are more likely to favor action A, which is better for industry interests. This is because researchers who are receiving lots of funding and producing lots of papers will tend to place more students in positions of influence. Over time, more and more scientists end up favoring action A over action B, even though action B is objectively superior.

In this way, industrial groups can exert pressure on the community of scientists to produce more results favorable to industry. And they do it simply by increasing the amount of work produced by well-intentioned scientists who happen to be wrong. This occurs even though the idealized scientists in the Holman-Bruner model are not people but just lines of computer code and so cannot possibly be biased or corrupted by industry lucre.

As Holman and Bruner point out, the normal processes of science then exacerbate this process. Once scientists have produced a set of impressive results, they are more likely to get funding from governmental sources such as the National Science Foundation. (This is an academic version of the "Matthew effect.")[53] If industry is putting a finger on the scales by funding researchers it likes, and those researchers are thus more likely to gain funding from unbiased sources, the result is yet more science favoring industry interests.

Notice also that if industrial propagandists are present and using selective sharing, they will disproportionately share the results of those scientists whose methods favor action A. In this sense, selective sharing and industrial selection can produce a powerful synergy.

Industry artificially increases the productivity of researchers who happen to favor A, and then widely shares their results. They do this, again, without fraud or biased production.

The upshot is that when it comes to methylmercury, even though we have no reason to think the researchers from the University of Rochester were corrupted by coal industry funding, the EPRI likely still got its money's worth. Uncorrupt scientists can still be unwitting participants in a process that subverts science for industry interests.

Holman and Bruner describe another case in which the consequences of industrial selection were even more dire. In 1979, Harvard researcher Bernard Lown proposed the "arrhythmic suppression hypothesis"—the idea that the way to prevent deaths from heart attack was to suppress the heart arrhythmias known to precede heart attacks.[54] He pointed out, though, that it was by no means clear that arrhythmia suppression would have the desired effect, and when it came to medical therapies, he advocated studies that would use patient death rates as the tested variable, rather than simply the suppression of arrhythmia, for this reason.

But not all medical researchers agreed with Lown's cautious approach. Both the University of Pennsylvania's Joel Morganroth and Stanford's Robert Winkle instead used the suppression of arrhythmia as a trial endpoint to test the efficacy of drugs aimed at preventing heart attacks.[55] This was a particularly convenient measure, since it would take only a short time to assess whether a new drug was suppressing arrhythmia, compared with the years necessary to test a drug's efficacy in preventing heart attack deaths. These researchers and others received funding from pharmaceutical companies to study antiarrhythmic drugs, with much success. Their studies formed the basis for a new medical practice of prescribing antiarrhythmic drugs for people at risk of heart attack.

The problem was that far from preventing heart attack death, antiarrhythmics had the opposite effect. It has since been estimated that their usage may have caused hundreds of thousands of premature deaths.[56] The Cardiac Arrhythmia Suppression Trial, conducted by the National Heart, Lung, and Blood Institute, began in 1986 and, unlike previous studies, used premature death as the endpoint. The testing of antiarrhythmics in this trial actually had to be discontinued ahead of schedule because of the significant increase in the death rate among participants assigned to take them.[57]

This case is a situation in which pharmaceutical companies were able to shape medical research to their own ends—the production and sale of antiarrhythmic drugs—without having to bias researchers. Instead, they simply funded the researchers whose methods worked in their favor. When Robert Winkle, who originally favored arrhythmia suppression as a trial endpoint, began to study antiarrhythmic drugs' effects on heart attack deaths, his funding was cut off.[58]

Notice that, unlike the Tobacco Strategy, industrial selection does not simply interfere with the public's understanding of science. Instead, industrial selection disrupts the workings of the scientific community itself. This is especially worrying, because when industry succeeds in this sort of propaganda, there is no bastion of correct belief.

While industrial selection is a particularly subtle and effective way to intervene directly on scientific communities, Holman and Bruner point out in an earlier article that industry can also successfully manipulate beliefs within a scientific community if it manages to buy off researchers who are willing to produce straightforwardly biased science.[59] In these models, one member of the scientific network is a propagandist in disguise whose results are themselves biased. For example, the probability that action B succeeds might be .7 for real scientists but only .4 for the propagandist. Figure 13 shows the structure of this sort of community.

An embedded propagandist of this sort can permanently prevent

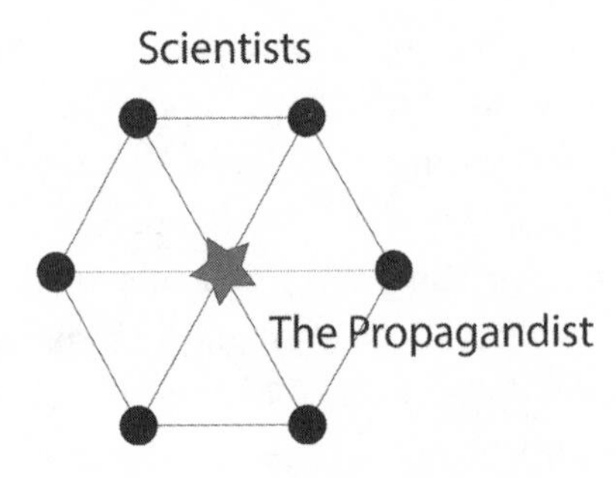

Figure 13. The structure of a model in which the propagandist directly shares biased research with scientists. Notice that unlike the network structures in figures 9, 10, and 12, the propagandist here does not focus on policy-maker belief, but poses as a scientist to directly sway consensus within the scientific community.

scientists from ever reaching a correct consensus. They do so by taking advantage of precisely the network structure that, as we saw in Chapter 2, can under many circumstances help a community of scientists converge to a true consensus. When the propagandist consistently shares misleading data, they bias the sample that generic scientists in the network update on. Although unbiased scientists' results favoring B tend to drive their credences up, the propagandist's results favoring A simultaneously drive them down, leading to indefinite uncertainty about the truth. In a case like this, there is no need for industry to distort the way results are transmitted to the public because scientists themselves remain deeply confused.

One important issue that Holman and Bruner discuss is how other scientists in the network can come to recognize a propagandist at work. They find that by looking at the distributions of their own results and those of their neighbors, scientists can under some circumstances identify agents whose results are consistently outliers and begin to discount those agents' results. Unfortunately, this is a difficult and time-consuming process—and it takes for granted that there are not too many propagandists in one's network. On the other hand, it highlights an important moral. In Chapter 2, we considered a model in which scientists chose whom to trust on the basis of their beliefs, under the assumption that successful scientists would share their own beliefs. But Holman and Bruner's work suggests that a different approach, though more difficult to implement in practice, might be more effective: namely, choose whom to trust on the basis of the evidence they produce.

The propaganda strategies we have discussed so far all involve the manipulation of evidence. Either a propagandist biases the total evidence on which we make judgments by amplifying and promoting results that support their agenda; or they do so by funding scientists whose methods have been found to produce industry-friendly results—which ultimately amounts to the same thing. In the most extreme case, propagandists can bias the available evidence by producing their own research and then suppressing results that are unfavorable to their position. In all of these cases, the propagandist is effective precisely because they can shape the evidence we use to form our beliefs—and thus manipulate our actions.

It is perhaps surprising how effective these strategies can be. They can succeed *without* manipulating any individual scientist's methods or results, by biasing the way evidence is shared with the public, biasing the distribution of scientists in the network, or biasing the evidence seen by scientists. This subtle manipulation works because in cases where the problem we are trying to solve is difficult, individual studies, no matter how well-conducted, tend to support both sides, and it is the overall body of evidence that ultimately matters.

But manipulating the evidence we use is not the only way to manipulate our behavior. For instance, propagandists can play on our emotions, as advertising often does. Poignancy, nostalgia, joy, guilt, and even patriotism are all tools for manipulation that have nothing to do with evidence.

The famous Marlboro Man advertising campaign, for instance, involved dramatic images of cowboys wrangling cattle and staring off into the wide open spaces of the American West. The images make a certain kind of man want to buy Marlboro cigarettes on emotional grounds. The television show *Mad Men* explored these emotional pleas and the way they created contemporary Western culture. Still, these sorts of tools lie very close to the surface. While it may be hard to avoid emotional manipulation, there is no great mystery to how it works. Rather than discuss these sorts of effects, we want to draw attention to a more insidious set of tools available to the propagandist.

One of Bernays's principal insights, both in his books and in his own advertising and public relations campaigns, was that trust and authority play crucial roles in shaping consumers' actions and beliefs. This means that members of society whose positions grant them special authority—scientists, physicians, clergy—can be particularly influential. Bernays argued that one can and should capitalize on this influence.

During the 1920s, Bernays ran a campaign for the Beech-Nut Packing Company, which wanted to increase its sales of bacon. According to Bernays, Americans had tended to eat light breakfasts—coffee, a pastry or roll, perhaps some juice. In seeking to change this, he invented the notion of the "American breakfast" as bacon and eggs. As he writes in *Propaganda:*

> The newer salesmanship, understanding the group structure of society and principles of mass psychology, would first ask: "Who is it that influences the eating habits of the world?" The answer, obviously, is: "The physicians." The new salesman will then suggest to physicians to say publicly that it is wholesome

> to eat bacon. He knows as a mathematical certainty, that large numbers of persons will follow the advice of their doctors.[60]

Bernays reports that he found a physician who was prepared to say that a "hearty" breakfast, including bacon, was healthier than a light breakfast. He persuaded this physician to sign a letter sent to thousands of other physicians, asking them whether they concurred with his judgment. Most did—a fact Bernays then shared with newspapers around the country.

There was no evidence to support the claim that bacon is in fact beneficial—and it is not clear that the survey Bernays conducted was in any way scientific. We do not even know what percentage of the physicians he contacted actually agreed with the assertion. But that was of no concern: what mattered was that the strategy moved rashers. Many tobacco firms ran similar campaigns through the 1940s and into the 1950s, claiming that some cigarettes were healthier than others or that physicians preferred one brand over others. There was no evidence to support these claims either.

Of course, the influence of scientific and medical authority cuts both ways. If the right scientific claims can help sales, the wrong ones can decimate an industry—as we saw earlier when the appearance of the ozone hole soon led to a global ban on CFCs. In such cases, a public relations campaign has little choice but to *undermine* the authority of scientists or doctors—either by invoking other research, real or imaginary, that creates a sense of controversy or by directly assaulting the scientists via accusations of bias or illegitimacy.

We emphasized in Chapter 1 that the perception of authority is not the right reason to pay attention to the best available science. Ultimately, what we care about is taking action, both individual or collective, that is informed by the best available evidence and therefore most likely to realize our desired ends. Under ideal circumstances, invoking—or undermining—the authority of science or medicine should not make any difference. What should matter is the evidence.

Of course, our circumstances are nowhere near ideal. Most of us are underinformed and would struggle to understand any given scientific study in full detail. And as we pointed out in the last chapter, there are many cases in which even scientists should evaluate evidence with careful attention to its source. We are forced to rely on experts.

But this role of judgment and authority in evaluating evidence has a dark side. The harder it becomes for us to identify reliable sources of evidence, the more likely we are to form beliefs on spurious grounds. For precisely this reason, the authority of science and the reputations both of individual scientists and of science as an enterprise are prime targets for propagandists.

Roger Revelle was one of the most distinguished oceanographers of the twentieth century.[61] During World War II, he served in the Navy, eventually rising to the rank of commander and director of the Office of Naval Research—a scientific arm of the Navy that Revelle helped create. He oversaw the first tests of atomic bombs following the end of World War II, at Bikini Atoll in 1946. In 1950, Revelle became director of the Scripps Institute of Oceanography.

In 1957, he and his Scripps colleague Hans Suess published what was probably the most influential article of their careers.[62] It concerned the rate at which carbon dioxide is absorbed into the ocean.

Physicists had recognized since the mid-nineteenth century that carbon dioxide is what we now call a "greenhouse gas": it absorbs infrared light. This means it can trap heat near the earth's surface, which in turn raises surface temperatures. You have likely experienced precisely this effect firsthand, if you have ever compared the experience of spending an evening in a dry, desert environment with an evening in a humid environment. In dry places, the temperature drops quickly when the sun goes down, but not in areas of high humidity. Likewise, without greenhouse gases in our atmosphere, the earth would be far colder, with average surface temperatures of about 0 degrees Fahrenheit (or −18 degrees Celsius).

When Revelle and Suess were writing, there had already been half a century of work on the hypothesis—originating with the Swedish Nobel laureate Svante Arrhenius and the American geologist T. C. Chamberlin[63]—that the amount of carbon dioxide in the atmosphere was directly correlated with global temperature and that variations in atmospheric carbon dioxide explained climactic shifts such as ice ages. A British steam engineer named Guy Callendar had even proposed that carbon dioxide produced by human activity, emitted in large and exponentially growing quantities since the mid-nineteenth century, was contributing to an increase in the earth's surface temperature.

But in 1957 most scientists were not worried about global warming. It was widely believed that the carbon dioxide introduced by human activity would be absorbed by the ocean, minimizing the change in atmospheric carbon dioxide—and global temperature. It was this claim that Revelle and Suess refuted in their article.

Using new methods for measuring the amounts of different kinds of carbon in different materials, Revelle and Suess estimated how long it took for carbon dioxide to be absorbed by the oceans. They found that the gas would persist in the atmosphere longer than most other scientists had calculated. They *also* found that as the ocean absorbed more carbon dioxide, its ability to hold the carbon dioxide would degrade, causing it to evaporate out at higher rates. When they combined these results, they realized that carbon dioxide levels would steadily rise over time, even if rates of emissions stayed constant. Things would only get worse if emissions rates continued to increase—as indeed they have done over the sixty years since the Revelle and Suess article appeared.

This work gave scientists good reasons to doubt their complacency about greenhouse gases. But just as important was Revelle's activism, beginning around the time he wrote the article. He helped create a program on Atmospheric Carbon Dioxide at Scripps and hired a chemist named Charles David Keeling to lead it. Later, Revelle helped Keeling get funding to collect systematic data concerning atmospheric carbon dioxide levels. Keeling showed that average carbon dioxide levels were steadily increasing—just as Revelle and Suess had predicted—and that the rate of increase was strongly correlated with the rate at which carbon dioxide was being released into the atmosphere by human activity.

In 1965, Revelle moved to Harvard. There he encountered a young undergraduate named Al Gore, who took a course from Revelle during his senior year and was inspired to take action on climate change. Gore went on to become a US congressman and later a senator. Following an unsuccessful presidential run in 1988, he wrote a book, *Earth in the Balance*, in which he attributed to Revelle his conviction that the global climate was deeply sensitive to human activity. The book was published in 1992, a few weeks before Gore accepted the Democratic nomination for vice president.

Gore's book helped make environmental issues central to the election. And he distinguished himself as an effective and outspoken advocate for better environmental policy. Those who wished to combat Gore's message could hardly hope to change Gore's mind, and as a vice presidential candidate, he could not be silenced. Instead, they adopted a different strategy—one that went through Revelle.

In February 1990, Revelle gave a lecture at the annual meeting of the American Association for the Advancement of Science, the world's largest general scientific society. The session in which he spoke was specifically devoted to policy issues related to climate change, and Revelle's talk was about how the effects of global warming might be mitigated.[64] Afterward, it seems that Fred Singer, whose service on the Acid Rain Review Panel we described in Chapter 1, approached Revelle and asked whether he would be interested in coauthoring an article based on the talk.

The details of what happened next are controversial and have been the subject of numerous contradictory op-eds and articles, and at least one libel suit.[65] But this much is clear. In 1991, an article appeared in the inaugural issue of a journal called *Cosmos*, listing Singer as first author and Revelle as a coauthor. The article asserted (with original emphasis), "We can sum up our conclusions in a simple message: *The scientific base for a greenhouse warming is too uncertain to justify drastic action at this time.*"[66] (If this sounds identical to Singer's message on acid rain, that is because it was.)

What was much less clear was whether Revelle truly endorsed this claim, which in many ways contradicted his life's work. (Revelle never had a chance to set the record straight: he died on July 15, 1991, shortly after the article appeared in print.)

It is certainly true that Revelle did not write the quoted sentence. What *Cosmos* published was an expanded version of a paper Singer had previously published, as sole author, in the journal *Environmental Science and Technology;* whole sentences and paragraphs of the *Cosmos* article were reproduced nearly word for word from the earlier piece. Among the passages that were lifted verbatim was the one quoted above.

Singer claimed that Revelle had been a full coauthor, contributing ideas to the final manuscript and endorsing the message. But others disagreed. Both Revelle's personal secretary and his long-term research assistant claimed that Revelle had been reluctant to be involved and that he contributed almost nothing to the text. And they argued that when the article was finalized, Revelle was weak following a recent heart surgery—implying that Singer had taken advantage of him.[67] (Singer sued Revelle's research assistant, Justin Lancaster, for libel over these statements. The suit was settled in 1994, with Lancaster forced to retract his claim that Revelle was not a coauthor. In 2006, after a ten-year period during which he was not permitted to comment under the settlement, Lancaster retracted his retraction and issued a statement on his personal website in which he "fully rescind[ed] and repudiate[d] [his] 1994 retraction." Singer told his own version of the story, which disagreed with Lancaster's in crucial respects, in a 2003 essay titled "The Revelle-Gore Story.")

Ultimately, though, what Revelle believed did not matter. The fact that his name appeared on the article was enough to undermine Gore's environmental agenda. In July 1992, *New Republic* journalist Gregg Easterbrook cited the *Cosmos* article, writing, "*Earth in the Balance* does not mention that before his death last year, Revelle published a paper that concludes: 'The scientific base for a greenhouse warming is too uncertain to justify drastic action at this time.'"[68] A few months later, the conservative commentator George Will wrote essentially the same thing in the *Washington Post.*

It was a devastating objection: it seemed that Revelle, Gore's own expert of choice, explicitly disavowed Gore's position.

Admiral James Stockdale—running mate of Reform Party candidate Ross Perot—later took up the issue during the vice presidential debate. "I read where Senator Gore's mentor had disagreed with some of the scientific data that is in his book. How do you respond to those criticisms of that sort?" he asked Gore.[69] Gore tried to respond—first over laughter from the audience, but then, when he claimed Revelle had "had his remarks taken completely out of context just before he died," to boos and jeers. He was made to look foolish, and his environmental activism naïve.

What happened to Gore was a weaponization of reputation. The real reason to be concerned about greenhouse gases has nothing to do with Roger Revelle or his opinion. One should be concerned because there is strong evidence that carbon dioxide levels are rapidly rising in the atmosphere, because increased carbon dioxide leads to dramatic changes in global climate, and because there will be (indeed there already are) enormous human costs if greenhouse gas emissions continue. There is still uncertainty about the details of what will happen or when—but that uncertainty goes in both directions. The chance is just as good that we have grossly *underestimated* the costs of global warming as that we have overestimated them. (Recall how scientists underestimated the dangers of CFCs.)

More, although the conclusion of the *Cosmos* article was regularly quoted, no evidence to support that conclusion was discussed by Easterbrook or Will in their articles. Indeed, the article offered no novel arguments at all. If Revelle had devastating new evidence that led him to change his mind about global warming, surely that should have been presented. But it was not.

But Gore himself had elevated Revelle's status by basing his environmentalism on Revelle's authority. This gave Singer—along with Will, Stockdale, and the many others who subsequently quoted the *Cosmos* article—new grounds for attacking Gore. Indeed, anyone who tended to *agree* with Gore was particularly vulnerable to this sort of argument, since it is precisely them who would have given special credibility to Revelle's opinion.

The details of how Singer and others used Revelle's reputation to amplify their message may seem like a special case. But this extreme case shows most clearly a pattern that has played a persistent role in the history of industrial propaganda in science.[70] It shows that how we change our beliefs in light of evidence depends on the reputation of the evidence's source. The propagandist's message is most effective when it comes from voices we think we can trust.

Using the reputations of scientists was an essential part of the Tobacco Strategy. Industry executives sought to staff the TIRC with eminent scientists. They hired a distinguished geneticist named C. C. Little to run it, precisely because his scientific credentials gave their activities stature and credibility. Likewise, they established an "independent" board of scientific advisors that included respected experts. These efforts were intended to make the TIRC look respectable and to make its proindustry message more palatable. This is yet another reason why selective sharing can be more effective than biased production—or even industrial selection. The more independence a researcher has from industry, the more authority he or she seems to have.

Even when would-be propagandists are not independent, there are advantages in presenting themselves as if they were. For instance, in 2009 Fred Singer, in collaboration with the Heartland Institute, a conservative think tank, established a group called the Nongovernmental International Panel on Climate Change (NIPCC). The NIPCC is Singer's answer to the UN's Intergovernmental Panel on Climate Change (IPCC). In 2007 the IPCC (with Gore) won the Nobel Peace Prize for its work in systematically reviewing the enormous literature on climate change and establishing a concrete consensus assessment of the science.[71]

The NIPCC produces reports modeled exactly on the IPCC's reports: the same size and length, the same formatting—and, of course, reaching precisely the opposite conclusions. The IPCC is a distinguished international collaboration that includes the world's most renowned climate scientists. The NIPCC looks superficially the same, but of course has nothing like the IPCC's stature. Such efforts surely mislead some people—including journalists who are looking for the "other side" of a story about a politically sensitive topic.

It is not hard to see through something as blatant as the NIPCC. On the other hand, when truly distinguished scientists turn to political advocacy, their reputations give them great power. Recall, for instance, that the founders of the Marshall Institute—mentioned in Chapter 1—included Nierenberg, who had taken over as director of the Scripps Institute after Revelle moved to Harvard; Robert Jastrow, the founding director of NASA's Goddard Institute for Space Studies; and Frederick Seitz, the former president of both the National Academy of Sciences and Rockefeller University, the premier biomedical research institution in the United States.

These scientists truly had made major contributions to their respective fields, and their reputations rightly put them in positions to exert influence even in areas where they had far less expertise. It was Nierenberg's status as former director of Scripps and a member of the National Academy of Sciences, as we saw in Chapter 1, that qualified him to serve as chair of the Acid Rain Peer Review Panel—and then gave him the opportunity to modify its executive statement.

These men's reputations also put them in a special position to criticize other scientists.[72] Perhaps the most striking example came in the aftermath of the second Assessment Report of the IPCC (not the NIPCC!), published in 1995.[73] This report included, for the first time, a chapter devoted to what is known as "fingerprinting," a set of methods for distinguishing climate change caused by human activity from that produced by sources such as sun cycles or volcanic activity. The chapter on this topic was written by a collaboration of distinguished scientists, but the "convening lead author," responsible for collecting the material together and orchestrating the whole chapter, was an American climate scientist named Ben Santer.

Santer was relatively junior when he took on this position—

though he had already made major contributions to fingerprinting methods and in many ways was ideally situated to convene that chapter. After the report was published, however, Seitz and others went on the attack. In an editorial published in the *Wall Street Journal*, Seitz accused Santer of violating scientific protocol by changing the final report to "deceive policy makers." But while it is true that Santer orchestrated late revisions to the chapter, they were made at the direction of IPCC chairman Bert Bolin, in response to comments from peer reviewers. Rather than a violation, they were in fact mandated by scientific protocol.

The Seitz letter reads as what it was: one of the most distinguished physicists in the United States publicly taking a much more junior scholar to task for impropriety: "The report's lead author, Benjamin D. Santer," Seitz writes, "must presumably take the major responsibility."[74] Bolin and forty other scientists signed a letter refuting Seitz's claims and asserting that Santer's work had been appropriate, to which Seitz and Singer both replied by more or less repeating the original accusations, but now with the implication that the whole IPCC was in on it.

Perhaps the group of forty internationally recognized climate scientists who signed the Bolin response defending Santer were enough to bolster Santer's reputation. But even if that is right, the end result is a display of authority on both sides, suggesting that there was a real controversy—not only on the facts, but on the scientific method itself—when in fact there was none. As the tobacco industry has shown, merely creating the appearance of controversy is often all the propagandist needs to do.

The weaponization of reputation is deeply connected to the polarization models we discussed in the last chapter. Recall that in those models, the key factor was trust. It was the people whose opinions we trusted who could exert the most influence over us; and as we came to trust some people more and others less, groups began to diverge, leading to polarized outcomes.

The proxy we used for trust was difference in belief on the matter at hand. This captures an important part of how polarization comes about—and it also explains some aspects of how reputation can be useful to the propagandist. After all, Revelle was admired by many people precisely because of his influential work on climate change. And it was at least in part because those people tended to agree with him on that topic that they were prepared to take any evidence he offered very seriously. (At least, the fact that *Gore* agreed with him suggests that Gore himself should take Revelle's opinion seriously.) When Easterbrook and Will implied that new evidence had influenced Revelle, that suggested it should also influence those who agreed with him.

But trust based on agreement about the topic at hand does not capture some of the other ways reputation can be weaponized. Trust often depends on other qualities—such as past behavior, personal connections, or professional training. Surely a lot of individual psychology is involved, too, in whom we trust and why. Even so, we can use our models to capture some interesting and important aspects of the weaponization of reputation by looking at the relationship between trust, belief, and scientific success.

In the models of polarization from the last chapter, everyone is trying to solve *one* problem and using credences about that problem to decide whom to trust. But in most real-world cases, we have more to go on than just the problem at hand. Suppose we have a network of scientists who are trying to solve two problems, instead of just one. For one of these problems, they are trying to choose between actions A and B; for the other, unrelated problem, they need to choose between actions Y and Z. Now suppose that for each problem, when deciding whom to trust, they consider their beliefs about *both* problems. The basic dynamics of the model are the same as before: each scientist uses Jeffrey's rule, which, remember, specifies how to update beliefs when you are not certain about evidence, but now the uncertainty they assign to a person's evidence depends on the distance between their beliefs on both topics.

In this model, we find that groups polarize, just as they do when only one problem is at stake. But now, perhaps surprisingly, when they polarize, they tend to form subcommunities whose beliefs on both problems are highly correlated. One finds, say, a subcommunity whose members all hold A and Z and another whose members all hold B and Y.[75] Note that the true beliefs do not necessarily correlate with one another: the model often yields cases in which the two communities each hold one true and one false belief.

These models suggest that one way to influence the opinions of members of a group is to find someone who already agrees with them on *other* topics and have that person share evidence that supports your preferred position. The idea is that people are relying on scientists' success on other problems to judge their general reliability. In other words, a scientist might think: "I am not so sure about actions A and B, but I am certain that action Z is better than Y. If another scientist shares my opinion on action Z, I will also trust that person's evidence on actions A and B."[76]

These sorts of effects can help to explain how weaponized reputation sometimes works. We look to people who have been successful in solving other problems and trust them more when evaluating their evidence. Established scientists with distinguished careers can presumably point to many past successes in correctly evaluating evidence. More, their peers—other scientists—have evaluated their past work and deemed it strong and reliable. When the established scientists present new evidence and arguments—even on topics that are completely unrelated to the field in which they established their reputations—you have good reason to trust them. In fact, philosophers of science Jeffrey Barrett, Brian Skyrms, and Aydin Mohseni have used network epistemology models to show that paying attention to this sort of past reliability can drastically improve the accuracy of a community's beliefs, on the assumption, of course, that no one is using reputation as a weapon.[77] These sorts of results can also help explain why someone like Seitz can be influential on a range of issues, including climate change, that are quite far from that person's previous research.

These issues of trust, on the problem at hand and more generally, surely play an important role in explaining how reputation gives some people outsized influence—at least within parts of a community. But as we will see, it is not the whole story. Conformity and network structure are also critical to understanding the social role of reputation in propaganda.

Lady Mary Wortley Montagu, born into the British aristocracy in 1689, was known for her brilliance and beauty.[78] (The London Kit-Cat Club, whose membership included the philosopher John Locke, once named her their beauty of the season.)[79] At the age of twenty-five, though, she suffered a bout of smallpox. She was lucky to live through it: at the time, smallpox killed 20–60 percent of those who were infected—including Lady Mary's brother, just two years before.[80] But the scarring ruined her legendary beauty.

Shortly thereafter, Lady Mary traveled to Turkey when her husband, Edward Wortley Montagu, was appointed as the British ambassador. She found herself delighted by the Turks' cultural practices: elaborate shopping centers, rooms spread wall to wall with Persian carpets, bathhouses where naked women drank coffee and juice. None of this could be found in Britain.

During her travels, Lady Mary encountered another strange practice: smallpox variolation. A version of modern-day inoculation, this typically involved scratching a person's arm and rubbing a scab, or fluid, from a smallpox pustule into the wound. Although a small percentage of people died from the resulting infection, the vast majority experienced a very mild form of smallpox and subsequently developed immunity.

Her enchantment with Turkish culture combined with her personal experience of smallpox made Lady Mary especially receptive to the practice. With the help of an English surgeon, Charles Maitland, and an old Turkish nurse, she successfully variolated her own young son. Upon returning home, in 1721 she began to advocate variolation in Britain. But she encountered a great deal of resistance from doctors. To many, the practice seemed strange, even barbaric. Perhaps just as bad was the fact that a woman was advocating a treatment developed by other women, and foreigners no less. Even Maitland, who had been happy to help perform the treatment in Turkey, was nervous to do so under the eyes of his English physician peers.[81] A debate raged in the London papers over whether variolation was safe and effective. The stakes were very high. At that very moment, a smallpox outbreak was claiming lives all over England.

The solution devised by Lady Mary was brilliant—and succeeded precisely because it appealed to the same conformist tendencies that had made physicians suspicious of variolation in the first place. She aimed to show that one of the most revered people in England was willing to have the practice performed on her own children. In 1722, at Lady Mary's behest, Caroline of Ansbach—then Princess of Wales, married to the Crown Prince of England—had her two daughters variolated. This demonstration provided direct evidence of the safety of variolation, but perhaps more important, it showed the citizenry of England that they would be in good company should they adopt it. The practice spread quickly among the English nobility, especially those with personal connections to Lady Mary or Princess Caroline.

The spread of variolation had little to do with new knowledge about its success or safety. Instead, it was shunned and then adopted because of social pressures. It was social pressures among doctors that at first prevented people from adopting it, and later, it was social pressures among nobility to share the beliefs and practices of the princess and her friends that accelerated its spread.

Although in this episode, social pressures propagated a true belief, the mechanisms by which Lady Mary persuaded her compatriots reflects an important way in which propagandists can take advantage of social networks. Remember the role of conformity in belief and behavior that we described in the last chapter. People generally prefer to conform their beliefs and actions to those around them—and this tendency can radically change the way information flows through communities. Those who hold true beliefs but who, from fear of social censure, do not share evidence supporting those beliefs, stop the spread of good ideas. This is what happened with Semmelweis, and it is what initially slowed variolation among the English.

We also argued in Chapter 2, however, that the influence of conformity on communities of scientists depends on the structure of the network of connections among those scientists. One such network structure was the wheel, wherein one individual is connected to many others, who in turn connect back to that central person. Real social networks do not exactly resemble wheels, but they often have substructures that look like the star in the wheel's center. Some individuals are connected to disproportionately many people and therefore have outsized social influence. Their actions tend to affect what the many people will do.

There is also a close connection here to some of the examples of the manipulation of reputation we described above. Princess Caroline was not a distinguished scientist, but she lived in a time when the inherent superiority of nobles, especially royalty, was taken as a given. It was precisely because of her reputation as a woman of great social standing that so many other members of the British nobility sought to be connected to her, and to conform their actions to hers.

A similar kind of conformity is at work in at least some of the cases of scientific reputation described above. It was on the basis of their reputations that figures such as Seitz were in a position to write articles and editorials that other people would read and listen to, and thus to reach many people who were relatively disconnected from one another. Although Seitz did not have a personal connection to all of his readers, the role of an opinion piece in a major paper is to place someone at the center of a star, at least briefly.

Likewise, Revelle had developed a large following, with many acolytes, including the future vice president of the United States, paying careful attention to his opinions. In part his reputation was based on his scholarly work and his demonstrated ability to gather and evaluate evidence. But surely the desire to be like him, to have an influence on the development of environmental science and policy, also played a role.

Propagandists who target those at the center of social stars can exploit our human tendencies to conform. Each time they persuade one major influencer to adopt a belief or practice, they can depend on them to persuade others who seek to emulate them. The genius behind Lady Mary's strategy was that she did not run around England indiscriminately telling every Martha and George about variolation. Instead, she targeted the wealthiest, most influential, and best-connected. She found the centers of the stars and got those people to adopt the new practice, wisely thinking that once that happened, variolation would spread on its own.

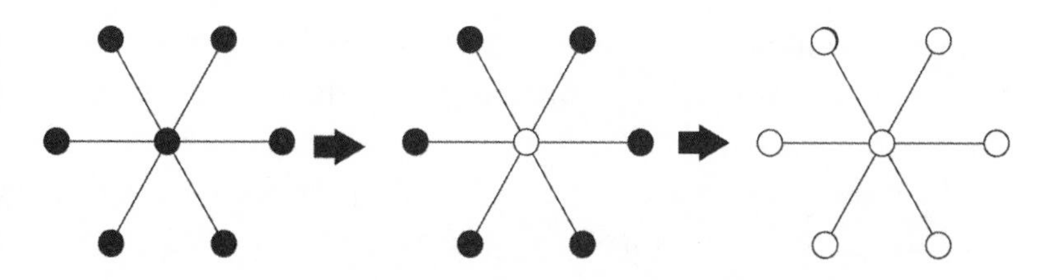

Figure 14. The spread of a belief in a star network. Dark nodes represent one action, and light nodes, another. Notice that the central individual is connected to many agents who are not connected to each other. This central individual thus holds special social influence over the rest of the network.

Figure 14 shows, in a highly simplified way, how something like this might unfold. In a network with uniform beliefs, if a central individual changes belief, that person exerts strong conformist influence on peripheral individuals, who will likely also change their beliefs.[82]

Targeting influential people to spread a new practice or belief is just one way propagandists take advantage of our conformist tendencies. The modern-day analogue to variolation—vaccination—remains a paradigmatic case in which conformist influences are at play. As in Lady Mary's day, there continue to be people—"anti-vaxxers"—who question the safety and wisdom of inoculating their children against disease. Strikingly, these people tend to cluster in neighborhoods, both physical and social, in which the discomforts of disagreement over a controversial topic can be avoided.[83] In some wealthy California neighborhoods, for example, vaccination rates have fallen as low as 20 percent, even though the state average is more than 85 percent.[84]

In 2017, a tight-knit Somali-American community in Minnesota experienced the state's worst measles outbreak since the 1980s.[85] After learning that rates of severe autism were particularly high in this group, anti-vaccine advocates posted fliers and ads throughout the Somali community center cautioning against vaccination. They also distributed pamphlets at community health meetings. Andrew Wakefield, the scientist who infamously, and falsely, first reported a link between vaccines and autism, visited Minneapolis repeatedly in 2010 and 2011 to talk with Somali parents of autistic children.[86]

As a result of these efforts, vaccination rates in the community dropped from 92 percent in 2004 to 42 percent in 2014, precipitating the measles outbreak. Even after the outbreak, the Vaccine Safety Council of Minnesota, an anti-vaccine organization, doubled down, hosting events and spreading misinformation about MMR vaccines and autism. There have even been reports of anti-vaxxers going door to door to try to persuade people that the measles outbreak was actually caused by the Health Department, whose goal was to pressure Somalis into vaccinating their children!

What we see here is anti-vaxxers peddling their beliefs—beliefs completely unsupported by evidence—to a close-knit group particularly susceptible to their message. In doing so, they take advantage of conformity effects. The Somali community in Minnesota is a classic case of a highly connected clique embedded in a larger group. By spreading the same information to peers who are especially likely to then discuss the matter and reinforce the messages they are getting, anti-vaxxers tap into this social network structure. If multiple members of a peer group can be brought around to a new way of thinking at once, conformity's usual resistance to change is instead co-opted to encourage the new practice. And once many members of the group decide not to vaccinate, the social effects within the group make it much more stable than one bold person bucking a trend.[87]

Figure 15 shows what this might look like. If a few members of the group change practices, there is pressure on the rest to change, and once they all agree, conformity should keep the whole group there.

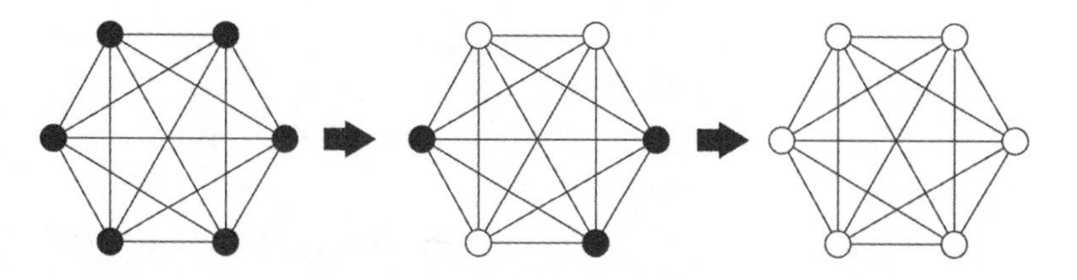

Figure 15. The spread of belief in a complete, that is, fully connected, network with conformity. Dark nodes represent one action, and light nodes, another. When multiple individuals in this network change actions, there is pressure on the rest of the group to change as well because of their desire to conform with neighbors. Once changed, the group is relatively stable.

In this chapter, we have looked at various ways in which propagandists can exploit how we share information and learn from one another to manipulate our beliefs and actions. The very same mechanisms that we saw in Chapter 2, which, remember, could both help us discover truth but also sometimes lead us astray, turn out to be levers for influencing our beliefs.

We have argued that there are multiple ways in which propagandists can succeed in this sort of manipulation. These can be as crude as producing scientific results that show precisely what you would like them to, or subtle, such as taking advantage of a scientific community's tendency to occasionally produce spurious results, and then sharing those results—or selecting those scientists who are most likely to produce spurious results and favoring them with extra funding.

We also argued that the propagandist can exert influence by targeting members of a community who, for one reason or another, have special influence. In the cases of Roger Revelle, Ben Santer, and others, part of this method involved identifying figures who were especially trusted by the community. For instance, scientists whose record of success on other problems has been very high may persuade others who have been influenced by these scientists' earlier work. A related method is to use conformity as a weapon by targeting individuals with particularly connected positions in the social network, or targeting cliques.

So far we have focused (mostly) on scientific communities and on the spread of information from scientists to others. But the same effects that influence scientists' beliefs also operate, often in dramatic ways, in society at large. In the next chapter we expand the picture we have been presenting to try to understand the influence of social factors on day-to-day beliefs.

FOUR

The Social Network

On Sunday, December 4, 2016, a twenty-eight-year-old man named Edgar Maddison Welch told his wife and two daughters that he had some business to attend to and left the house.[1] He got into his car and drove about six hours from his home in Salisbury, North Carolina, to a pizzeria in Washington, D.C. He carried with him an AR-15 semiautomatic rifle, a handgun, and a shotgun, all loaded. When he arrived at the restaurant at about 3 p.m., he entered carrying the AR-15 and opened fire, unloading multiple rounds into a locked door inside the establishment.

Welch thought he was a hero. He believed that the pizzeria, known as Comet Ping Pong, was the staging ground for an international child prostitution ring headed by none other than Hillary Clinton, the former Democratic nominee for president. Welch was there to investigate the pizza parlor—and if possible, save the children.

A little over a month earlier, shortly before the 2016 election, FBI director James Comey had announced that he was reopening an investigation into Hillary Clinton's use of a private email server while secretary of state. (Recall that it was the fact that Comey had previously closed this investigation, without recommending prosecution, that, according to ETF News, had prompted Pope Francis to endorse Donald Trump for president.) The reason for Comey's announcement was the presence of an unknown number of possibly new emails on a computer that had been confiscated from the home of a top Clinton aide. The circumstances were sordid: the computer belonged to disgraced former congressman Anthony Weiner, who had recently been accused of sending nude pictures of himself to a fifteen-year-old girl. (Weiner eventually pleaded guilty to transferring obscene material to a minor.) Weiner's wife, Huma Abedin, was a top Clinton aide.

Two days later, a post on the social media site Twitter alleged that the emails in fact revealed something far worse than the original allegation that classified material had passed through Clinton's server. Citing "NYPD [New York Police Department] sources," the tweet claimed that the emails implicated Hillary Clinton as the kingpin of "an international child enslavement and sex ring."

This tweet was shared more than six thousand times during the next week.[2] The following day, an article appeared on the website YourNewsWire.com claiming that an "FBI insider" had confirmed allegations of a pedophile sex ring linked to many people in the US government, including several sitting members of Congress and, of course, Hillary Clinton. The story, quickly shared on other blogs and news aggregators, became the topic of multiple new fake news stories. Some of these were verbatim reproductions of the YourNews Wire.com article, while others included new text, new allegations, and new claims of inside sources. One website, called Subject: Politics, claimed that the NYPD had "raided" a Clinton property and found further damaging material. (No such raid occurred.)

On the heels of these articles, online sleuths began investigating other publicly available materials—including a slew of emails that were apparently stolen from Clinton campaign chair (and former Bill Clinton chief of staff) John Podesta and released on the website of the organization WikiLeaks.[3] These amateur investigators soon came to believe that they had detected a code hidden in Podesta's lunch orders. Starting from "cheese pizza," which has the same initials as "child prostitution," they created a translation manual: "hotdog" meant "boy," "pasta" meant "little boy," "sauce" meant "orgy," and so on.[4] A discussion board soon appeared on the website Reddit with the title "Pizzagate," where these allegations were discussed and new "evidence" was brought to bear; other discussions continued on websites popular with right-wing youths, such as 4chan.

In a particularly baffling inference, this community somehow reasoned that since pizza-related terms were involved in their "code," the prostitution ring must be run out of a pizzeria. (Yes, a real pizzeria—that one was not code!) Soon they identified a location: D.C.'s Comet Ping Pong, which was owned by a man with links to the Clintons (he had previously dated the CEO of pro-Clinton media watchdog Media Matters for America) and which Podesta was known to frequent. (Their investigation also determined that Podesta practiced witchcraft and drank the blood of his human victims.)

These bizarre and evidence-free allegations soon spread beyond the dark underbelly of the internet to relatively mainstream right-wing media such as the Drudge Report and Infowars. Infowars is far from a reliable source of information, but it has an enormous following. Alex Jones, its creator and principal voice, has more than 2 million follows on YouTube and 730,000 followers on Twitter; by spreading the rumors, Jones vastly increased their reach. (He later apologized for his role in the whole affair.)[5] Soon the pizzeria was flooded with phone calls threatening violence, and hundreds of obscene restaurant "reviews" appeared online, echoing the allegations.

By the end of November, the story had gotten so much attention that mainstream news sources, including the *New York Times* and the *Washington Post*, ran articles debunking the claims—which, predictably, only led to more attention for the conspiracy theory, along with articles and videos attempting to refute the debunkers.[6]

Welch became aware of the theories around this time. He had first been told the stories by friends in North Carolina; but in the days before he drove to D.C., he had had an internet service installed in his house—at which point he was able to read about the Pizzagate allegations for himself. What he found was deeply concerning to him. He got the "impression something nefarious was happening"—and that no one else was doing anything about it.[7] He decided to take matters into his own hands.

Our focus in the previous three chapters has been on science. As we explain in the Introduction, the reason for this is that it is relatively clear that in scientific communities some or all of the actors involved are trying to learn about the world in what they take to be the most reliable and effective ways possible. They *want* to discover the truth, in the sense discussed in Chapter 1. They want to discover what events actually happened and their physical causes. The members of these communities can certainly be wrong, and often are. But there is no doubt that they are in the business of gathering and evaluating evidence. This makes it particularly clear how our models apply—and all the more significant that there are so many ways for these communities to end up mistaken.

But as we also argued in the Introduction, science can be thought of as an extreme case of something we are all trying to do in our daily lives. Most of us are not trained as scientists, and even fewer have jobs in which they are paid to do research. But we are often trying to figure stuff out about the world—and to do this, we use the same basic kinds of reasoning that scientists do. We learn from our experience—and, crucially, we learn from the experiences of others.

Inasmuch as all of us are learning from the world and from one another, our models of communities of scientists apply just as well to groups of ordinary people. The mechanisms we have identified for the spread of specific beliefs operate in the population at large in similar ways. And understanding these mechanisms and how they can be subverted to other people's ends tells us a great deal about the political situation today in the United States, the United Kingdom, and much of Western Europe.

The American public, for example, is deeply divided on many issues. They include some of the topics discussed in this book, such as global climate change. But Americans are also polarized on whether the Affordable Care Act—aka Obamacare—should be the basis for future healthcare policy or repealed and replaced by a completely different policy; whether the multilateral treaty through which Iran has agreed to give up its nuclear weapons program should be abandoned; whether free-trade agreements ultimately improve the country's economic conditions; how tightly the government should (or legally may) regulate guns; and whether lowering corporate tax rates and tax rates on high earners will stimulate middle-class wage growth.

Virtually all of these cases feature disagreements about basic matters of fact that contribute to the disagreements about policy. These disagreements themselves arise because people tend to trust different information sources: some rely on MSNBC, the *New York Times*, or the *Washington Post*, whereas others look to Fox News, the *Wall Street Journal*, and the *Washington Times.* Some point to studies produced by the Heritage Foundation, the Cato Institute, or the Heartland Institute, whereas others look to the Southern Poverty Law Center, the Brookings Institution, or the Center for American Progress. Some point to less dependable sources such as Breitbart

News, Infowars, AlterNet, and the Palmer Report—or even worse, RT or Sputnik.

When friends, family members, colleagues, and especially strangers disagree with our views, it is easy to attribute this disagreement to the failures of those people: they are ignorant of the facts, too emotional to properly evaluate the situation, or too stupid. But what if that is not what's going on? Or at least, what if ignorance and emotion are only part of the story—and perhaps not the most important part?

Emotion plays no role in our models. Neither does intelligence nor political ideology. We have only very simple, highly idealized agents trying to learn about their worlds using (mostly) rational methods. And they often fail. Moreover, they can be readily manipulated to fail, simply by an agent taking advantage of the same social mechanisms that, in other contexts, help them to succeed.

What if these sorts of social factors lie behind the spread of "fake news" and even the bleeding of conspiracy theories into mainstream sources such as the *Washington Post* and Fox News?

"Fake news" has a long history, particularly in the United States. In the decades immediately before and after the American Revolution, for instance, partisans on all sides attacked their opponents through vicious pamphlets that were often filled with highly questionable accusations and downright lies. Likewise, fake news arguably launched the Spanish American War.[8] After the USS *Maine*—a US warship sent to Havana in 1898 to protect American interests while Cuba revolted against Spain—mysteriously exploded in Havana Harbor, several US newspapers, most notably the *New York World* and *New York Journal*, began to run sensational articles blaming Spain for the explosion and demanding a war of revenge.[9] (The actual cause of the explosion was and remains controversial, but concrete evidence has never been produced that Spain was involved.)[10] Ultimately, spurred in part by pressure from the news media, the US government gave Spain an ultimatum that it surrender Cuba or face war—to which the Spanish responded by declaring war on the United States.[11] (Spain sued for peace fewer than three months later.)

In 1835, the *New York Sun*, a politically conservative but generally reputable newspaper, published a series of six articles asserting that the English astronomer John Herschel had discovered life on the moon.[12] The articles claimed to have been reprinted from an Edinburgh newspaper and contained a number of alleged quotes from Herschel. They even included illustrations of winged hominids Herschel was said to have seen. Needless to say, there is no life on the moon—and Herschel never claimed to have found it. The articles were never retracted. (Compare these claims to ones made by a guest on Alex Jones's Infowars radio show in June 2017 to the effect that NASA is running a child slave colony on Mars.)[13]

Nine years later, Edgar Allan Poe published a story in the *Sun* in which he described (as factual) a trans-Atlantic hot-air balloon journey by a famous balloonist named Monck Mason.[14] This, too, never occurred. (The article was retracted two days later.)

So fake news has been with us for a long time. And yet something has changed—gradually over the past decade, and then suddenly during the lead-up to the 2016 UK Brexit vote and US election.

In 1898, when the *New York World* and *New York Journal* began agitating for war, they had large circulations. The *New York Journal* claimed 1.25 million readers per day, allegedly the largest in the world.[15] (New York City then had a population of about 3.4 million; the *Journal* figures are surely inflated, but perhaps not by much: the aggregate circulation for New York dailies in 1900, according to the U.S. Census that year, was more than 2.7 million.) But their audiences consisted almost exclusively of New Yorkers—and not even all New Yorkers, as the better-respected *Times*, *Herald Tribune*, and *Sun* also had wide readerships. Regional newspapers outside New York generally did not pick up the *World* and *Journal* articles calling for war with Spain. Although the stories surely influenced public opinion and likely contributed to the march toward war, their impact was limited by Gilded Age media technology.

In the past decade, these limitations have vanished. In February 2016, Facebook reported that the 1.59 billion people active on its website are, on average, connected to one another by 3.59 degrees of separation. Moreover, this number is shrinking: in 2011, it was 3.74.[16] And the distribution is skewed, so that more people are *more* closely connected than the average value suggests. According to the Pew Research Center, 68 percent of American adults use Facebook (out of 79 percent of American adults who use the internet at all).[17] Twitter, meanwhile, has about 70 million users each month in the United States—a bit under 30 percent of American adults—and about 330 million users worldwide.[18] Information posted and widely shared on Facebook and Twitter has the capacity to reach huge proportions of the voting public in the United States and other Western democracies.

Even if fake news is not new, it can now spread as never before. This makes it far more dangerous. But does anyone actually believe the outrageous stories that get posted, shared, and liked on social media?

These stories' persistence could have other explanations. Perhaps some people find them funny or unbelievable, or share them ironically. Others may share them because, even though they know the content is false, the stories reflect their *feelings* about a topic. Many internet "memes"—digital artifacts that are widely shared on the internet—have the character of elaborate jokes, which often signal some sort of social status or engagement.[19] Is fake news the same?

Perhaps. But some people *do* believe fake news. Clearly Edgar Welch, for instance, believed that the Comet Ping Pong pizzeria harbored trafficked children. And he is not alone. In a survey conducted by the polling firm Ipsos Public Affairs for BuzzFeed News, 3,015 US adults were shown six headlines, three of which were real news and three of which were fake.[20] The headlines were drawn from a list of the most-shared news items in the months before the election, and in total, they had been shared a similar number of times. Respondents were asked whether they recalled the headlines and, for those they did recall, whether they were accurate.

One-third of the survey respondents recalled seeing at least one of the fake news headlines. Those who remembered the headline judged the fake news to be "very" or "somewhat" accurate 75 percent of the time. By contrast, respondents who recalled seeing the real news headlines judged them to be accurate 83 percent of the time.

Other surveys and experiments have found results broadly consistent with this picture.[21] A Pew survey of 1,002 adults found that 23 percent admitted to having shared fake news—of whom 73 percent admitted that they did so unwittingly, discovering only later that the news was fake.[22] (The others claimed to have known that it was fake at the time but shared it anyway.) Of course, these results do not include participants who unwittingly shared fake news and never learned that it was fake, nor do they include those who would not admit to having been duped.

There is a famous aphorism in journalism, often attributed to a nineteenth-century *New York Sun* editor, either John B. Bogart or Charles A. Dana: "If a dog bites a man it is not news, but if a man bites a dog it is." The industry takes these as words to live by: we rarely read about the planes that do not crash, the chemicals that do not harm us, the shareholder meetings that are uneventful, or the scientific studies that confirm widely held assumptions.

For many issues, focusing on novel or unexpected events is unproblematic. Novelty makes things salient, and salience sells papers and attracts clicks. It is what we care about. But for some subjects, including science as well as politics and economics, a novelty bias can be deeply problematic.

We saw in Chapter 3 that a key feature of the Tobacco Strategy was to amplify and promote research produced by legitimate, unbiased scientists that tended to support the tobacco industry's agenda. This was extremely effective, and in the models we have considered, simply amplifying a subset of the evidence by sharing it more broadly can lead a community of policy makers, or the public, to become confident in a false belief, even as scientists themselves converge to the true belief. The basic takeaway here is that when trying to solve a problem in which the evidence is probabilistic or statistical, it is essential to have a complete and unbiased sample. Focusing on only part of the available evidence is a good way to reach the wrong belief.

This sort of biasing can happen even if no one is actively trying to bias the evidence shared with the public. All it takes is a mechanism by which the evidence is selectively disseminated.

This is precisely what happens when journalists focus on novel, surprising, or contrarian stories—the sorts that are most likely to gain attention, arouse controversy, and be widely read or shared. When journalists share what they take to be most interesting—or of greatest interest to their readers—they can bias what the public sees in ways that ultimately mislead, even if they report only on real events.

To better understand how this sort of thing can happen, we looked at a variation of the propagandist models described in Chapter 3. Now, instead of having a group of policy makers connected to scientists and also connected to a propagandist, we imagine that we have a collection of policy makers who receive all of their information from a single third party: a journalist, who scours the scientific results and shares the evidence that she judges to be most novel.

We suppose that the journalist is updating her beliefs in light of all of the scientific evidence, so that as more evidence is gathered, she tends to converge to whatever the scientific consensus is. But she only *shares* what is most surprising. There are a few ways to implement this basic idea. For instance, one way is to have the journalist look at each round of scientists' results and then share whatever result or results from that round are the most unlikely. Another way is to have the journalist share all of the results that pass some kind of threshold, given her current beliefs: that is, the ones that the journalist thinks are sufficiently strange to be of interest.[23]

In both of these cases, we find that the public sometimes converges to the false belief, even when the journalist and the scientific community converge to the true one.[24]

In such a model, the journalist will generally share some results that are unlikely because they point so strongly in the direction of the true belief, and also some that are unlikely because they point strongly toward the false belief. She is not biasing the evidence in the explicit way that a propagandist would. Sometimes the journalist will actually *strengthen* our confidence in the true belief, relative to what the science supports. But that does not mean that her actions are neutral. As we saw in the propagandist examples, it is the total distribution of evidence that tends to lead us to the true belief. And intervening to change this distribution will change where consumers of that evidence end up.

There is another way in which journalists can unwittingly spread false beliefs. Journalism has legal and ethical frameworks that seek to promote "fairness" by representing all sides of a debate. From 1949 until 1987 the US Federal Communications Commission even maintained a policy called the Fairness Doctrine that *required* media with broadcast licenses to offer contrasting views on controversial topics of public interest. The rule no longer applies—and even if it did, few people get their news from broadcast media any longer. But since few journalists relish being accused of bias, pressures remain for journalists to present both sides of disagreements (or at least appear to).

Fairness sounds great in principle, but it is extremely disruptive to the public communication of complex issues. Consider again the model of a journalist selectively communicating results. Now, instead of the journalist sharing only surprising results, suppose that every time she chooses to share a result supporting one view, she also shares one supporting the other view—by searching through the history of recent results to find one that reflects the "other perspective."

In figure 16 we show a possible sample of outcomes for scientists performing trials (with ten data points each). The journalist shares the two bolded results, one randomly selected from those that support B and one randomly selected from those that support A. The policy makers will see two studies where B was successful six times and three times, whereas the true distribution of results—nine, six, six, and three—would lead to a much more favorable picture of B's general success.

What happens?[25] In this case, the policy makers do tend to converge to the true belief when the scientists do. But this convergence is substantially slower for policy makers than for the scientists—and indeed, it is substantially slower than if the journalist had merely shared two *random* results from the scientific community. This is because we generally expect evidence favoring the true belief to appear more often. Sharing equal proportions of results going in both directions puts a strong finger on the scale in the wrong direction. Indeed, norms of fairness have long been recognized as a tool for propagandists: the tobacco industry, for instance, often invoked

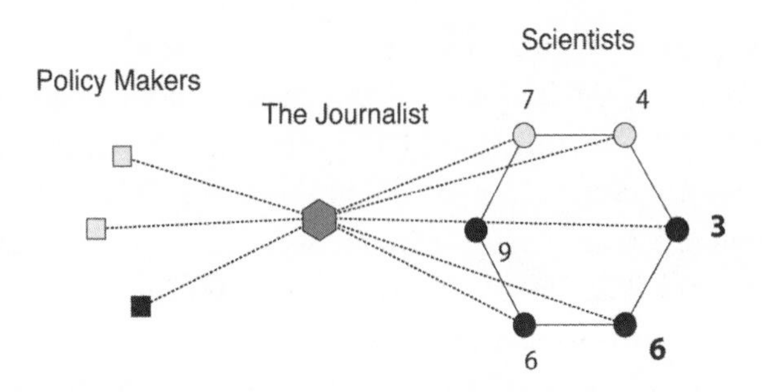

Figure 16. Communication in a model with a "fair" journalist who chooses two results to report to policy makers. Upon observing results from scientists, the journalist communicates one that supports theory B and one that spuriously supports A (bolded) to policy makers. This has the effect of biasing the set of data available to policy makers. Light nodes represent individuals who favor action A, and dark nodes, B.

the Fairness Doctrine to insist that its views be represented on television and in newspaper articles.[26]

Ultimately, the mere existence of contrarians is not a good reason to share their views or give them a platform. And the suggestion that it would be unfair *not* to report contrarian views—or at least, not to report them specifically *as* contrarian views—especially when the scientific community has otherwise reached consensus, is wrong, at least if what we care about is successful action. This is so even when the contrarians have excellent credentials and legitimate results that do support their views. When we are trying to solve difficult problems, there will always be high-quality and convincing evidence that pushes us in both directions.

You might worry about this: after all, throughout this book we have pointed to cases in which "consensus" views of experts have turned out to be false. What if journalists had reported only the views of those gentleman physicians who insisted that their hands could not possibly transmit disease, dismissing Semmelweis as a habitual contrarian? Indeed, something very much like this occurred in the lead-up to the Iraq War in 2003: foreign-policy experts in the United States and the United Kingdom, along with politicians across the political spectrum, almost uniformly adopted the view that Saddam Hussein was developing weapons of mass destruction. (He was not.) The Bush and Blair administrations used these alleged weapons to justify launching a war that ultimately led to more than a decade of severe civil strife and nearly two hundred thousand civilian deaths, according to the Iraq Body Count project.[27] The *New York Times* was widely criticized for presenting this consensus without adequate scrutiny or skepticism, and the editors took the highly unusual step of issuing an apology in 2004.[28] In this case, reporting only the consensus view and stories that were broadly in line with it had dire consequences.

Fair enough. So how, then, are journalists to tell the difference—especially when they are not experts in the relevant material? For one, stories that are ultimately about current or historical events have a very different status from stories about science. It is not, and should not be, journalists' role to referee scientific disagreements; that is what peer review and the scientific process are for, precisely because expert judgment is often essential. It is in those cases in which that process has already unfolded in the pages of scientific journals, and the losers of the debates wish to rehash them in public forums, that journalists should be most cautious. On the other hand, it most certainly *is* journalists' job to investigate and question those purported matters of fact on which major domestic and foreign policies are based—including determining whether there is a scientific consensus on an issue relevant to policy-making.

Perhaps more important, it is essential to focus on the reasons for apparent controversy. We do not wish to rule out the possibility that today's Semmelweis is laboring in obscurity, or that a talented reporter could not find him and bring his ideas to public attention.

This, after all, is precisely what *20/20* did with Hightower and methylmercury in fish. Our point, rather, is that the mere *existence* of contrarians or (apparent) controversy is not itself a story, nor does it justify equal time for all parties to a disagreement. And the publication of a surprising or contrary-to-expectation research article is not in and of itself newsworthy.

So we need journalists to avoid sensationalizing new findings and to report both that there is a consensus (when there is one) and the reasons for it. It is particularly important for journalists and politicians to carefully vet the sources of their information. It will invariably be the case that nonexperts need experts to aggregate evidence for them. This is what propagandists seek to exploit, by standing in for disinterested experts and aggregating evidence in a way favorable to their own interests. Often the groups doing this aggregation consciously attempt to mislead journalists about their independence and credentials.

This is where institutions can play an important role. Journalists reporting on science need to rely not on individual scientists (even when they are well-credentialed or respected), but on the consensus views of established, independent organizations, such as the National Academy of Sciences, and on international bodies, such as the Intergovernmental Panel on Climate Change. Likewise for other matters of fact important to policy-making, such as tax and economic policy. In nations with reliably independent government record keepers—such as the Bureau of Labor Statistics in the United States, which measures economic indicators such as unemployment and inflation, or the Congressional Budget Office, which scores proposed legislation on expected budgetary impact—the findings of these organizations should be given special weight, as compared with self-appointed experts claiming that the "official" figures are faulty, misleading, or biased. Reports from the United Nations, particularly when they involve serious peer review, as with the IPCC, are also often more reliable than those from the governments of individual nations. Such institutions can certainly be manipulated for partisan ends,[29] but they are far more likely to be reliable than individuals or organizations whose interests are tied to the issues at stake.

In the predawn hours of July 10, 2016, a man named Seth Rich called his girlfriend as he walked home from a Washington, D.C., bar.[30] The bar was only a few miles away from his home, but Rich walked slowly and they chatted for a few hours. And then he abruptly ended the call. Moments later, at about 4:20 a.m., Rich was shot twice in the back. Police responded immediately and took him to a local hospital, but he soon died of his wounds.

Rich was twenty-seven years old. Just a few days earlier, he had accepted a job as a staffer for Hillary Clinton's presidential campaign. The Democratic National Convention, at which Clinton would be officially nominated as the party's presidential candidate, would open in about two weeks. For the previous two years, Rich had worked for the Democratic National Committee (DNC), tasked with increasing voter turnout during the contentious primary between Clinton and Bernie Sanders.[31]

The D.C. Metropolitan Police ruled the death a homicide, a robbery attempt gone wrong. But within a few days of Rich's shooting, conspiracy theories began to swirl, apparently initiated by a post on the internet discussion site Reddit.[32] The allegation was that Rich had been killed by the Clintons, or perhaps Debbie Wasserman Schultz, then head of the DNC, to cover up election fraud during the primary campaigns. Commentators on the thread speculated that through his job at the DNC, Rich had learned of efforts to suppress Sanders voters or otherwise tilt the primary toward Clinton, and that this information had made him dangerous to the Clinton campaign.

Rumors that a Clinton had murdered someone who presented a political liability were hardly new. In 1994, William E. Dannemeyer, a former Republican congressman from Southern California who had been voted out of office two years previously, sent a letter to congressional leaders asserting that then-President Bill Clinton had been responsible for dozens of murders in his rise to the presidency.[33] Dannemeyer listed twenty-four people connected with the Clintons who he claimed had died "under other than natural circumstances." His list appears to have originated with an activist named Linda Thompson, who the previous year had produced thirty-four names of people who she believed—admittedly with "no direct evidence"—had been killed by Clinton or his associates. A 1994 film called *The Clinton Chronicles* made similar charges.[34]

These claims were investigated at length by law enforcement and various special prosecutors during Bill Clinton's presidency. They have no substance. And yet they have persisted among conspiracy theorists on both the far right and the far left, creating an ecosystem in which the allegations concerning Rich seemed unsurprising, even plausible.

Soon the story took on new texture. About a month before Rich was killed, the DNC admitted that its computer servers had been hacked and thousands of emails stolen.[35] The first indications of a hack had come nine months previously, when the FBI contacted the DNC to say that at least one of its computers had been compromised by Russian operatives. A technician checked the system and found nothing amiss, and the DNC did not respond further. Two months later, the FBI again contacted the DNC, this time to say that one of its servers was routinely sending data to Russia. Once again, the DNC's IT staff decided that the computers had not been breached and did not communicate the FBI's warnings to the DNC leadership. The DNC apparently did nothing to stop the hack.

That Russian hackers had compromised DNC computers was not widely reported until June 14, 2016. The DNC acknowledged it and attributed it to Russian security services the following day. Meanwhile, WikiLeaks founder Julian Assange announced that he had thousands of pages of emails related to the Clinton campaign that he would soon release.[36] WikiLeaks waited another month to release the emails, apparently to maximize damage to Clinton's campaign: the messages appeared online on July 22, 2016, days before the Democratic convention opened. Some of the leaked emails suggested that the DNC had taken actions to support Clinton's candidacy and to undermine that of Sanders.

A hacker claiming to be Romanian and calling himself Guccifer 2.0 claimed responsibility for the hack. But the FBI, which launched an official investigation on July 25, 2016, indicated that it believed Russian hackers were behind the attacks—an accusation consistent with its earlier warnings to the DNC. Half a dozen private cybersecurity firms came to the same conclusion, and by October 2016, there was a broad consensus among the US intelligence agencies that the hacks had been conducted by Russian intelligence services.[37]

Still, WikiLeaks never confirmed Russian involvement. In August 2016, Assange implied that the messages had come from a different source altogether: Seth Rich. He offered a twenty-thousand-dollar reward for information related to Rich's death, and then, in an interview with the Dutch television program *Nieuwsuur*, he asserted, following a question about Rich, that "our sources take risks"—before adding that WikiLeaks has a policy of never revealing or otherwise commenting on its sources.[38]

The story's lack of evidence, or even coherence, did not stop some media outlets from reporting these allegations. Some of this coverage came from RT, the English-language Russian propaganda site, which seemed to like having Rich as an alternative culprit for the DNC hack.[39] But more mainstream sources also reported on Assange's remarks, including the *Washington Post*, BuzzFeed News, and *Newsweek*.

The story got a further boost in May 2017 when Fox News and several Fox affiliates ran stories in which it was claimed that the hacked DNC emails had been discovered, by the FBI, on Rich's computer.[40] These allegations were attributed to a man named Rod Wheeler, a former Washington, D.C., homicide detective who had been paid to work on the Rich case by a Republican insider. Wheeler talked as if he had himself seen the messages on Rich's computer and could speak directly to this new evidence.

There was only one problem: the Fox News story was completely fabricated. Shortly after it appeared, the FBI stated that it was not involved in the Rich investigation. Soon Wheeler admitted that he had not, after all, seen the emails on Rich's computer. Rich's parents produced a statement saying that they had not seen or heard of any evidence that Rich had ever possessed or transmitted any DNC emails—or that Rich's death was anything other than a botched robbery. Either the evidence did not exist, or whoever had it was withholding it.

About a week after first airing the stories, Fox and its affiliates retracted them; that August, Wheeler filed a lawsuit against Fox claiming that its reporters had knowingly fabricated quotations they attributed to him and that the entire story had been orchestrated in consultation with the White House.[41] It seems that there is not, and never has been, any reason to believe that Rich had any involvement in the DNC email hack. Yet at least some right-wing media personalities, including Fox's own Sean Hannity, have continued to repeat variations on these claims. Hannity, in particular, has refused to issue a retraction over the Wheeler remarks—even after Wheeler himself disavowed them.

The Rich example shows how thin the line between "real news" and "fake news" can be. Of course, Fox News (like ETF News—and MSNBC) has a transparent political orientation; but by running a story based on the remarks attributed to Wheeler, it veered from editorial slant into blatant falsehood. The example shows that even legitimate news sources can produce and spread fake news. And if the allegations in Wheeler's lawsuit that the White House was involved in the story are true, the situation is even more troubling.

Some readers will surely respond that Fox News, especially the commentary side that includes Hannity, is not a legitimate news source at all. But it is not the only mainstream source to have spread fake news since the 2016 election. MSNBC host Stephanie Ruhle, for instance, claimed on air that Fox News had held its 2016 Christmas parties at the Trump International Hotel in Washington, D.C. The claim was false, and she later apologized.[42] (The Republican National Committee, however, does seem to have held a Christmas party at the hotel that year.)[43] And CNN Investigates, an elite CNN reporting team, was forced to retract two stories related to Trump during the summer of 2017: in one, they claimed that Anthony Scaramucci, who very briefly served as White House communications director, had been connected to a Russian investment fund; in another, they claimed that former FBI director James Comey would offer testimony to Congress, which he never ultimately gave.[44]

Fox News is in a different category from ETF News or other right-wing sources such as Breitbart or Infowars. So are CNN and MSNBC. What *makes* them different is that Fox, CNN, and MSNBC generally retract false stories in light of new evidence. A self-correcting editorial process is at work.[45] One can quibble about how long it takes to make the retractions, and whether whatever damage was going to be done had already happened. But in the cases where the facts are simply and demonstrably wrong, these media sources have corrected the errors—and in many cases, they have done so because other news sources have policed them.[46] That these groups can nonetheless spread falsehood makes the problem of identifying fake news much more difficult. It is not as simple as pointing to reliable and unreliable sources.

These sorts of blatant falsehoods, though, are not the only problem, and maybe not the worst one. A deeper issue concerns a more subtle way in which fake news shades into real news: it sets a journalistic agenda.[47] Unlike in the Pizzagate fiasco, real facts exist in the background of the Rich story. Rich really was murdered, there really have been allegations that he was involved in the DNC email leak, WikiLeaks really did offer a large reward for information related to the case, and real investigations have been privately funded and conducted to identify his killers—who, by the way, have not been identified or arrested. It is not "fake" news to report on these facts, particularly given that a large readership wants to pore over such stories.

But it is also hard to see it as *real* news if the only reason anyone was interested in the first place was that these facts are tied up with widespread fake news. Ultimately, fake news, unsubstantiated allegations, and innuendo can create interest in a story that then justifies investigations and coverage by more reliable sources. Even when these further investigations show the original allegations to be baseless, they spread the reach of the story—and create the sense that there is something to it. Were it not for the conspiracy theories, Assange would never have been asked about Seth Rich, the *Washington Post* would not have covered his remarks (or refuted them), private investigators would not have been funded by Republican lobbyists for Fox News to quote, and so on.

Here is another manifestation of a theme that has come up throughout this book. Individual actions that, taken on their own, are justified, conducive to truth, and even rational, can have troubling consequences in a broader context. Individuals exercising judgment over whom to trust, and updating their beliefs in a way that is responsive to those judgments, ultimately contribute to polarization and the breakdown of fruitful exchanges. Journalists looking for true stories that will have wide interest and readership can ultimately spread misinformation. Stories in which every sentence is true and impeccably sourced can nonetheless contribute to fake news and false belief.

These dynamics are troubling, but once we recognize them, it appears that small interventions could have a significant impact. In particular, it is important to distinguish two essentially different tasks that reliable news sources perform. One involves investigating allegations, checking facts, and refuting false claims. This is an important activity—but it is also a risky one, because it can, counterintuitively, expand the reach of fake stories. In some cases, as with the Comet Ping Pong conspiracy, it can turn fake news into a real story. The other task involves identifying and reporting real stories of independent interest, relevant to readers' lives and responsive to their demands. This is also an important activity, and although it requires judgment, it runs fewer risks of promoting falsehoods.

We suggest that these activities need to be kept firmly separated—and that media which serve as primary news sources, such as the *Wall Street Journal*, *New York Times*, and *Washington Post*, should consider avoiding the first sort of activity altogether. Checking facts and policing the media, while extremely important, are best left to independent watchdogs such as PolitiFact and Snopes.com, which run less of a risk of driving further traffic and attention to false stories.

•

There is overwhelming evidence that foreign actors, apparently associated with Russian intelligence services, attempted to interfere in the 2016 US election—initially to weaken Hillary Clinton's candidacy, and later to promote Donald Trump.[48] As we noted, for instance, the US intelligence community has determined that Russian intelligence services hacked the DNC's email servers and then released embarrassing emails, particularly related to the treatment of Bernie Sanders.[49] Facebook, Twitter, and Google have revealed that accounts linked to the Russian government spent well over one hundred thousand dollars to purchase political ads, most of which seem to have been designed to create controversy and sow civil discord. Facebook has subsequently revealed that Russian-produced political content reached as many as 126 million US users.[50] As of this writing, allegations of explicit coordination have not been fully settled one way or the other, but there is little doubt that the Trump campaign knew that the Russian government favored its candidate and was making efforts to influence the election in his favor.

For our purposes, however, there are even more important ways in which Russia-backed groups appear to have influenced the election, and Western political discourse more generally.

For instance, let us simply stipulate for the sake of argument that Russian agents did in fact hack the DNC servers and release some or all of the emails they stole. What was their purpose in doing so? If gathering intelligence were the whole goal, it would make little sense to release the hacked emails. The character and scale of the release suggest a different motive. Ultimately, the release of the DNC emails has led to lasting divisions and sustained controversies within the Democratic Party, which in turn have affected Democrats' ability to effectively govern. In other words, the emails have produced discord and mistrust—and in doing so, they have eroded an American political institution. Perhaps this was the point.

The idea that Russia's goal was to create social division and undermine democratic (small *d*) institutions is supported by the ways Russia-backed groups appear to have used social media during the campaign. Although it seems they aimed to help Trump's candidacy, they did not target their efforts only at potential Trump voters or others on the political right. Russian organizations are reported to have developed a bestiary of personalities, voices, and positions crafted to influence various groups,[51] ranging from members of the LGBTQ community, to Black Lives Matter activists, to gun rights supporters, to anti-immigrant zealots, and even, bizarrely, to animal lovers.

One goal appears to have been to establish trust with a broad range of demographic and interest groups in order to influence them. The *New York Times* reported in October 2017 that the Russia-linked LGBT United Twitter account, for example, declared: "We speak for all fellow members of LGBT community across the nation. Gender preference does not define you. Your spirit defines you." Even the dog-lover page, the *Times* suggests, was probably formed with the intention of developing a set of followers who could then be slowly exposed to political content. In other words, the goal was to get close, pose as a peer, and then exert influence.

After convincing users that they shared core beliefs and values, Russians used these platforms to widen the gap between right and left. Their "Heart of Texas" Facebook page tried to persuade Texans to secede from the nation. Their "Blacktivist" group agitated for protests in Baltimore in response to the death of Freddie Gray, a twenty-five-year-old black man who died shortly after being transported without safety restraints in a police van. One bizarre campaign included a contest titled "Don't Shoot Us," which encouraged Pokémon Go users to capture Pokémon near real-world sites of police brutality and name them after the victims.[52]

In none of these cases were group members pushed to accept

views far from those they already held. The LGBTQ group, for instance, did not advocate for Trump; instead, social influence was used to push people to more extreme versions of the views they already held.

The picture that emerges is one in which Russian propagandists were highly sensitive to the dynamics of social influence, in several ways. First, recall the role that network structure plays in the models of conformity we presented in Chapter 2: agents in some positions in a network are much more influential than those in other positions. For instance, in Chapter 3 we observed that the agent at the center of a star or wheel structure is more influential than those on the periphery. In part this is simply a matter of having a larger overall number of connections to other agents, but that is not the whole story. It also matters that the other agents are more weakly connected to one another than the central agent is to everyone.

At least some of the Facebook content linked to Russian accounts were "pages" or "community pages" rather than "groups." It is easy to use the terms indistinguishably, but the distinction matters: a Facebook "group" is primarily designed to facilitate discussions among its members. A Facebook "page" is designed for an organization or celebrity to create a community of their followers. "Community pages" are somewhere in between: there, other users can sign up as "followers" of the page and can make posts directed to the community; but whoever created the page can make posts as well, which are then targeted at and shared with the whole community.

If you visit one of the Russian-linked community pages—say, the LGBT United page, which was active as of March 2018—the first things you see are the posts created by the group itself, identified as "LGBT United" posts; the posts directed at the group from other members of the community can be found only by following a further series of links. In other words, the community page mimics a star network, with the page creator at the center. Rather than try to influence people already at the center of star networks, the Russian accounts appear to have *created* star networks by creating affinity groups structured so that they could communicate to the whole community more easily than members could communicate with one another.

Of course, the mere ability to broadcast information is not sufficient to create influence. You also need those to whom you are broadcasting to listen. And here we see the value of creating and distributing content through groups defined by a subject of shared interest or agreement—be it the right to bear arms or the right to love kittens.

Recall that the central assumption underlying our models of polarization was that people are more susceptible to the influence of those they trust—in particular, those who they think have formed reliable beliefs in the past. This means that establishing connections through affinity groups provides powerful tools for influence, especially when the influence tends to push them farther in directions they are already inclined to go. And if the purpose is merely to drive polarization—as opposed to persuading everyone of any particular claim—posing to people on both sides of an issue as someone who shares their opinions, and then presenting further evidence or arguments in support of those opinions, will be very successful.

Structuring these interactions around affinity groups may have allowed the Russia-linked groups to exert even more influence. Recall the modification to the polarization models that we described in Chapter 3, when we discussed reputation. There we considered the role that shared beliefs in one domain could play in influencing beliefs in other domains. We described how, in a model in which each agent considers agreement on a range of issues when determining how much to trust another agent on any particular issue, we found that different beliefs influenced one another: populations would become polarized on multiple different beliefs but would do so in a strongly correlated way.

How might a propagandist use this fact to manipulate beliefs? In the examples we considered in Chapter 3, the strategy was to find someone who had demonstrated success in solving other problems—someone with a long track record of discovering the truth—and have that person act as a mouthpiece for the propagandist.

But another strategy is available. If we consider beliefs across a range of issues in determining whom to trust, then establishing the existence of shared beliefs in one arena (opinions on gun laws or LGBTQ rights) provides grounds for trust on issues in other arenas. So if someone wanted to convince you of something you are currently uncertain about, perhaps the best way for that person to do it is to first establish a commonality—a shared interest or belief. The affinity groups appear to have played just this role.

One natural response to fake news is to say that social media sites, web search providers, and news aggregators have a responsibility to identify fake news and stop it. It was, after all, a Facebook algorithm that actively *promoted* a fake story, hosted on ETF News, about Megyn Kelly. In other cases, these sites merely provided vehicles for fake news stories to go viral. This is in part a matter of what people are interested in reading about—but it is amplified by algorithms that social media sites use to identify which stories are highly engaging so as to maximize their viewership.

We fully endorse this solution: organizations like Facebook, Twitter, and Google *are* responsible for the rampant spread of fake news on their platforms over the past several years—and, ultimately, for the political, economic, and human costs that resulted. But assigning culpability is not the same as identifying solutions.

If it was algorithms on social media sites that amplified and spread fake news, one might hope that algorithms could also help to identify fake news and *prevent* it from being amplified on these sites—or even prevent fake news from being shared on such sites at all. But many sites that produce or distribute fabricated stories—including ETF News—also produce stories that are true. Should it be possible to share those true stories if they come from unreliable sources? And as we have just seen, media groups that typically produce stories that are true or essentially true can also sometimes produce false stories. Would those stories make it through the filter? If so, how many fake, or at least false, news stories should a given site be able to run before it gets flagged as a fake news provider?

Since the 2016 election, a large number of academic articles have appeared that offer algorithmic solutions to the fake news problem.[53] These efforts are surely worthwhile, but determining what is true is a difficult, time-consuming, and human-labor-intensive process. (The Kelly story was promoted by Facebook days after Facebook fired its human editors.) Algorithmic responses can help, but more is needed: ultimately, we need human editorial discretion, armies of fact checkers, and ideally, full financial and political independence between the groups whose actions are covered by news organizations, whose platforms are used to distribute news broadly, and who are responsible for evaluating whether claims are true. We need to recognize fake news as a profound problem that requires accountability and investment to solve.

Perhaps more important, we need to recognize that fake news stories—and propaganda more generally—are not fixed targets. These problems cannot be solved once and for all. Economist Charles Goodhart is known for "Goodhart's law," which has been glossed by anthropologist Marilyn Strathern as, "When a measure becomes a target, it ceases to be a good measure."[54] In other words, whenever there are interests that would like to game an instrument of measurement, they will surely figure out how to do it—and once they do, the measurement is useless. A classic example occurred in Hanoi, Vietnam, under French colonial rule. In the early 1900s, the city was so overrun with rats that the colonial government offered a bounty for every rat tail returned to them. The problem was that people began cutting off rats' tails and then simply releasing the rats to breed and make more rats, with more tails.[55]

We should expect a similar response from fake news sources. As soon as we develop algorithms that identify and block fake news sites, the creators of these sites will have a tremendous incentive to find creative ways to outwit the detectors. Whatever barriers we erect against the forces of propaganda will immediately become targets for these sources to overcome. Bennett Holman, for example, uses historical cases to illustrate how pharmaceutical companies constantly devise new strategies to get around reforms by consumer protection groups, prompting further action on the part of these groups, and so on. He compares the process to an asymmetric arms race.[56]

The term "arms race" will conjure, for most people, the nuclear arms race between the US and the USSR during the Cold War, where each side attempted to out-arm the other, leading to the proliferation of ever more dangerous and massive bombs. In biology, we see arms races between species with conflicting interests: cheetahs get faster to catch gazelles, who in turn evolve to outrun cheetahs; prey species concentrate neurotoxins while their predators evolve an ever-increasing resistance to the poisons.

This framework paints a dreary picture of our hopes for defeating fake news. The better we get at detecting and stopping it, the better we should expect propagandists to get at producing and disseminating it. That said, the only solution is to keep trying.

The idea that our search for truth in public discourse is an endless arms race between highly motivated, well-funded political and industrial forces attempting to protect or advance their interests, and a society trying to adapt to an ever-changing media and technological landscape, suggests that would-be propagandists and others who would seek to distort the facts will constantly invent new methods for doing so. If we hope to have a just and democratic society whose policies are responsive to the available evidence, then we must pay attention to the changing character of propaganda and influence, and develop suitable responses.

The models and examples we have discussed suggest some interventions that could help us fight fake news—and propaganda more generally. Just as we have focused on only one aspect of why false beliefs spread—social effects rather than individual psychological considerations, which surely also matter—our proposals on how we might best respond, as a society, to fake news and propaganda will also be only part of the story. But we do think that recognizing the importance of social effects on the spread and persistence of false beliefs, even if we assume that all individuals in a society are perfectly rational (which, alas, they are not), shows that whatever else we do, we *also* need to think about interventions that take networks into account.

One possible intervention concerns the relationship between local issues and issues that are more abstract, in the sense of being disconnected from individuals' everyday experiences. The more local our politics is, the less chance for it to be dominated by distorting social effects of the sort that have emerged in recent years. This is because policies with local ramifications give the world more chances to push back. Consider the difference between national legislation aimed at regulating emissions generally, and local legislation aimed at regulating emissions from the nearby coal plant. Or legislation on mercury contamination in a town's fishing areas that has observable effects on the day-to-day lives of those who would vote on the bill. This generates a situation better matched by models in which there is a notable difference between the success rates of actions A and B, and where conformity is less significant. We should expect in such cases that social effects will not matter as much, and that it will be harder for industry to take advantage of these effects to obscure the truth.

Of course, some pressing issues, such as economic policy and climate policy, are necessarily national or international problems that cannot be made local in the same way. But the more individual politicians can do to redirect political discourse toward issues of local significance, the more we should expect voters to be guided by the best evidence available. Ironically, this suggests that one of the drivers of political dysfunction in the United States may have been anticorruption efforts related to earmark spending—that is, to federal appropriations used to support projects in individual congressional districts. (In 2011, President Obama declared in his State of the Union address that he would veto any bill that contained earmarks, and the following month, the Republican-controlled Congress enacted a ban on such spending.) Voters who are worried about real, tangible consequences to their lives may be less likely to engage in conformist identity signaling during elections.

A second possible intervention concerns our ability to construct social networks that minimize exposure to dissenting opinion and maximize positive feedback for holding certain beliefs, independent of their evidential support. Social media sites should change the algorithms by which information is shared among members so that all members of the site are exposed to material that is widely shared and discussed among subgroups (rather than limiting exposure to just those subgroups).[57] Remember that in our conformity models, dissenting opinions were bolstered and protected by cliquish, clumpy social networks. When individuals are mostly connected within one small group, outside influences matter less than those within the clique, and conformity effects can buffer the group from evidence that conflicts with their false beliefs. Even if some members gather this evidence, they are unwilling to share it for fear of social censure.

Simply sharing more information with all members of such a group may not disrupt their conformity, but the more people in a network with access to reliable information, the more likely that someone will manage to buck the social trends.

Of course, even if we could magically scramble all the social networks in the United States, this would not solve the issue of distrust. As we have discussed, even if every member of a group can communicate, the group can still become polarized when people do not trust those with different beliefs. Perhaps ironically, propagandists have demonstrated remarkably effective interventions for such situations. To persuade a group to change beliefs, you need to find someone who shares their other core beliefs and values, and get that person to advocate your view. We do not recommend setting up fake Facebook interest groups to persuade, say, anti-vaxxers to change their minds. But we do recommend finding spokespeople whose shared values can ground trust with groups that are highly dubious of well-established facts.

In an ideal world, trusted politicians might play this role. If Republican representatives who believe in anthropogenic climate change, for example, were willing to share this with their conservative constituents, they might have a serious impact on public opinion. In fact, joint work by Aydin Mohseni and Cole Williams suggests that when individuals go against the trends in their own social networks to hold a minority belief, their statements of this belief can be especially powerful. Because other people expect them to conform, it is easy to infer that they must have good reasons for taking such a socially risky position.[58] We might call this the "maverick effect"—such as when Arizona senator (and self-styled maverick) John McCain says that climate change is real, his statement has much more impact on the right than the same one made by Al Gore. This same mechanism, of course, worked in the opposite direction in the case of Roger Revelle.

The implication is that finding, targeting, and publicizing the views of a few individuals who are willing to go against the political and social consensus to spread true beliefs can have an outsized social effect. It is even better if these individuals are (like politicians) highly connected. Those at the center of a star are under unusual pressure to conform, that is, not to play the maverick, but they also have considerable power to sway their peers when they decide to do so.

We can keep hoping that politicians will do the right, rather than the expedient, thing. But there may be *more* hope for the rest of us.

One general takeaway from this book is that we should stop thinking that the "marketplace of ideas" can effectively sort fact from fiction.[59] In 1919, Justice Oliver Wendell Holmes dissented from the Supreme Court's decision in *Abrams v. United States* to uphold the Sedition Act of 1918.[60] The defendants had distributed leaflets denouncing US attempts to interfere in the Russian Revolution. While the court upheld their sentences, Holmes responded that "the ultimate good desired is better reached by free trade in ideas. . . . The best test of truth is the power of the thought to get itself accepted in the competition of the market."

Holmes's admirable goal was to protect free speech, but the metaphor of the marketplace of ideas, as an analogue to the free market in economics, has been widely adopted. Through discussion, one imagines, the wheat will be separated from the chaff, and the public will eventually adopt the best ideas and beliefs and discard the rest. Unfortunately, this marketplace is a fiction, and a dangerous one. We do not want to limit free speech, but we do want to strongly advocate that those in positions of power or influence see their speech for what it is—an exercise of power, capable of doing real harm. It is irresponsible to advocate for unsupported views, and doing so needs to be thought of as a moral wrong, not just a harmless addition to some kind of ideal "marketplace."

This is as true for scientists as it is for political and social leaders. Remember the propaganda models presented in Chapter 3. These showed that studies that erroneously support false beliefs are essential tools for propagandists. This is not the fault of scientists, but on the (certain) assumption that industry interests are here to stay, it is still incumbent on scientists to take whatever measures they can to prevent their work from being used to do social damage.[61]

This means, first, that scientific communities must adopt norms of publication that decrease the chances of spurious findings, especially in cases when the public good is clearly on the line. Second, scientists need to consider inherent risks when they publish. Philosopher of science Heather Douglas has argued persuasively that it is the responsibility of scientists to take into account, throughout the scientific process, the social consequences of the work they do and to weigh these against the benefits of publishing—or at least, to hold their own research on socially sensitive topics to particularly high standards before they do choose to publish it.[62] One might respond that the duties of scientists are just to do science. We side with Douglas in thinking that in areas where industrial propagandists are at work, the risks to society from publishing are sufficiently high that they must factor into scientists' decisions.[63]

There are other ways to do science that minimize the risks of playing into the hands of propagandists. Throughout the present book, we have pointed out that low-powered studies are especially likely to generate spurious results. One solution is for scientific communities to raise their standards. Another is for groups of scientists to band together, when public interest is on the line, and combine their results before publishing. This generates one extremely strong piece of research with a clear consensus, rather than many disjointed, weaker studies with disparate results. This strategy has the downside of suppressing contrary views, but that is also the point. If scientists work out disagreements ahead of time, they protect the public from those who would use dissent to sow confusion.[64]

Where it is impossible for scientists to work together in this way, independent bodies—possibly government organizations but ideally independent scholarly organizations such as the National Academy of Sciences—should oversee the aggregation and presentation of disciplinary consensuses. Such a step does not avoid the problem of propagandists who promote research that aligns with their interests, but it would make abundantly clear to anyone paying attention that the studies to which the propagandists want to direct our attention are not consistent with the state of the art.

Another clear message is that we must abandon industry funding of research. It is easy to think, as a scientist, that one is incorruptible. As we have discussed, though, even in cases in which scientists are not corrupt, industry funding can drastically alter the scientific process through industrial selection. Industry funding is tempting, since it is plentiful and research is expensive. But if we want good science that protects the interests of the public, this expense is a cost the public must bear. It is too costly to allow industry to interfere with science, and too easy for it to do so, even when scientists are individually trustworthy.

Journalists, to minimize the social spread of false belief, need to hold themselves to different standards when writing about science and expert opinion. As we have argued, attempts at fairness often bias the scientific evidence seen by the public. Giving a "fair shake" to minority viewpoints in science can grant authority and power to fringe elements or downright bad actors. Everything we have seen in this book indicates that journalists should instead try to give to the public an unbiased sampling of available evidence. If there are ninety-nine studies indicating that smoking is dangerous for every one study indicating the opposite, journalists should talk to ninety-nine scientists who think smoking is harmful for every one who does not. (John Oliver, on his satirical news show, recently did just this in real time, by bringing ninety-seven climate scientists and three skeptics onto his stage at once—or at least actors who did a good job of posing as such.)[65]

In this vein, Wikipedia has developed a commendable standard for writing about controversial scientific topics.[66] The Wikipedia "proper weighting" standard holds that if multiple opinions or views are expressed in peer-reviewed articles in journals that are indexed by reputable scientific databases, then it is appropriate to include all such opinions in a Wikipedia article. But the weight given to each such opinion—that is, the space allotted to the opinion relative to other such opinions—should be proportional both to the number of published articles in high-impact journals that support the view, and to the number of citations such articles have received, with more recent articles and citations given greater weight than older ones. This may sound like a complicated standard to meet, but in fact, modern scholarly tools (including, for instance, Google Scholar) can make it very easy to identify which articles are highly cited and which opinions are widely defended in reputable journals.

Of course, whatever respectable journalists do, their effect will be muted as long as other sources in the marketplace peddle false or misleading material. On this point, we have a controversial suggestion. We currently have a legislative framework that limits the ability of certain industries—tobacco and pharmaceuticals—to advertise their products and to spread misinformation. This is because there is a clear public health risk to allowing these industries to promote their products. We also have defamation and libel laws that prohibit certain forms of (inaccurate) claims about individuals. We think these legislative frameworks should be extended to cover more general efforts to spread misinformation. In an era of global warming, websites like Breitbart News and Infowars are more damaging to public health than Joe Camel and the Marlboro Man were in the past, and they should be treated as such.[67]

In many ways, the United States is behind Europe on this front. Like the US and UK, France, Germany, the Netherlands, and other European nations have been targeted by fake news—often with alleged links to Russia.[68] In 2015, the European Union created a specialized task force called East StratCom, with the express purpose of identifying strategies to combat fake news and propaganda.[69] (The group is called "East" StratCom because the EU recognized Russia as the source of troubling disinformation campaigns as early as 2015.) The group draws on a large network of volunteers to identify and debunk disinformation—a kind of institutionalized version of Snopes.com or PolitiFact. More recently, Germany has implemented new laws aimed at holding social media companies responsible for "unlawful" content that remains on their sites—including material deemed to be hate speech.[70] At the time of writing, French president Emmanuel Macron has endorsed a similar law against fake news more broadly.[71]

Some readers may consider this a form of censorship and counter to the spirit of free speech.[72] But the goal here is not to limit speech. It is to prevent speech from illegitimately posing as something it is not, and to prevent damaging propaganda from getting amplified on social media sites. If principles of free speech are compatible with laws against defamatory lies about individuals, surely they are also compatible with regulating damaging lies dressed up as reported fact on matters of public consequence. Lying media should be clearly labeled as such, for the same reason that we provide the number of calories on a package of Doritos or point out health effects on a cigarette box. And social media sites should remain vigilant about stopping the spread of fake news on their platforms or, at the very least, try to ensure that this "news" is clearly labeled as such.

We expect these suggestions, if implemented, to be just another step in the long arms race with propagandists. For this reason, part of the picture will have to involve regulatory bodies in government as well as online sources whose entire purpose is to identify and block sources of misinformation. This will require significant social resources to do well. But when the safety and well-being of nations and the world depend on it, it seems like the least we can do.

We conclude this book with what we expect will be the most controversial proposal of all. This suggestion goes beyond the core issues of truth, falsehood, science, and propaganda that we have focused on. We believe that any serious reflection on the social dynamics of false belief and propaganda raises an unsettling question:

Is it time to reimagine democracy?

We do not mean to express skepticism about the ideals of a democratic society—properly construed. (More on that in a moment.) But we do think that the political situation among Western democracies suggests that the institutions that have served us well—institutions such as a free and independent press, publicly funded education and scientific research, the selection of leaders and legislators via free elections, individual civil rights and liberties—may no longer be adequate to the goal of realizing democratic ideals.

In a pair of important books—*Science, Truth, and Democracy* (2001) and *Science in a Democratic Society* (2011)—the philosopher of science Philip Kitcher has presented a vision of what it means to do science in a way that is responsive to the needs of democracy—and also what it means to have a democracy that is suitably responsive to the facts uncovered by that science.

We wish to extract from Kitcherism an idea about what it means to have a democratic society that is responsive to fact. When it comes to decisions about and informed by science—which we may think of, broadly, as everything from clearly scientific subjects such as climate change to calculations of the actual flows of immigrants across the US-Mexico border—what he calls "vulgar democracy" is simply unacceptable. Vulgar democracy is the majority-rules picture of democracy, where we make decisions about what science to support, what constraints to place on it, and ultimately what policies to adopt in light of that science by putting them to a vote. The problem, he argues, is simple: most of the people voting have no idea what they are talking about. Vulgar democracy is a "tyranny of ignorance"—or, given what we have argued here, a tyranny of propaganda. Public beliefs are often worse than ignorant: they are actively misinformed and manipulated.[73]

As we have argued throughout this book, it is essential that our policy decisions be informed by the best available evidence. What this evidence says is simply not up for a vote.[74]

There is an obvious alternative to vulgar democracy that is equally unacceptable. Decisions about science and policy informed by science could be made by expert opinion alone, without input from those whose lives would be affected by the polices. As Kitcher points out, this would push onto scientific elites decisions that they are not qualified to make, because they, too, are substantially ignorant: not about the science, but about what matters to the people whose lives would be affected by policy based on that science.

Kitcher proposes a "well-ordered science" meant to navigate between vulgar democracy and technocracy in a way that rises to the ideals of democracy. Well-ordered science is the science we would have if decisions about research priorities, methodological protocols, and ethical constraints on science were made via considered and informed deliberation among ideal and representative citizens able to adequately communicate and understand both the relevant science and their own preferences, values, and priorities.[75] But as Kitcher is the first to admit, there is a strong dose of utopianism here: well-ordered science is what we get in an ideal society, free of the corrupting forces of self-interest, ignorance, and manipulation. The world we live in is far from this ideal. We may strive for well-ordered science, but it is not what we have.

As it stands, matters of crucial public interest—the habitability of huge swaths of our planet; the population's susceptibility to disease and its exposure to pollutants, toxins, and radiation—are decided in a way that mimics the mechanisms of vulgar democracy without realizing any of its ideals. Before it can influence policy, hard-won knowledge is filtered through a population that cannot evaluate it—and which is easily manipulated. There is no sense in which the people's preferences and values are well-represented by this system, and no sense in which it is responsive to facts. It is a caricature of democracy.

Of course, replacing this system with well-ordered science is beyond impractical. What we need to do instead is to recognize how badly our current institutions fail at even approximating well-ordered science and begin reinventing those institutions to better match the needs of a scientifically advanced, technologically sophisticated democracy: one that faces internal and external adversaries who are equally advanced and constantly evolving. We need to develop a practical and dynamic form of Kitcherism.

Proposing our own form of government is, of course, beyond the scope of this book. But we want to emphasize that that is the logical conclusion of the ideas we have discussed. And the first step in that process is to abandon the notion of a popular vote as the proper way to adjudicate issues that require expert knowledge.

The challenge is to find new mechanisms for aggregating *values* that capture the ideals of democracy, without holding us all hostage to ignorance and manipulation.

Notes

INTRODUCTION

1. This text has been translated into modern English; see, for example, Mandeville (1900), or, for more context, see Higgins (2011).
2. Odoric's journal was translated by Sir Henry Yule in 1866; see Odoric of Pordenone (2002). This episode is discussed by Lee (1887).
3. He did, however, allegedly encounter men with the heads of dogs and trees that produced bread.
4. Joannes Jonstonus described the horned hare in his 1658 tome *Historiae Naturalis de Quadrupedibus*. An illustration of it can be found in Joris Hoefnagel's *Animalia Quadrupedia et Reptilia*, published circa 1575. Duret (1605) devotes a chapter to zoophytes, plant-animal hybrids. A member of the Royal Society, Sir Robert Moray (1677), reported that he had investigated creatures called "barnacle geese," which allegedly grew on trees in Western Ireland, and had found tiny, perfect geese with beaks and wings inside the barnacles.
5. This history is drawn from Lee (1887).
6. The original article is still available online ("Pope Francis Shocks World" 2016).
7. See FBI National Press Office (2016).
8. Allcott and Gentzkow (2017) estimate an upper bound of twenty reads per Facebook engagement but cite other authors who have measured rates between three and fourteen read-throughs per engagement on Facebook and similar platforms.

9. Here we are following work by Craig Silverman (2016b) at BuzzFeed News, who conducted a detailed analysis of election-related stories shared on social media during the months leading up to the election.
10. A detailed report of the results of the 2016 presidential election is available at politico.com ("2016 Election Results: President Live Map by State" 2016) and at cnn.com ("2016 Election Results: State Maps" 2016).
11. See Allcott and Gentzkow (2017).
12. Likewise, during the period from May until the end of August, ETF News' top five articles garnered 1.2 million engagements on Facebook.
13. This was reported, for example, by the *Washington Post* (Ohlheiser 2016).
14. The history of WTOE 5 News is described by Silverman and Singer-Vine (2016), and the original article is refuted by Snopes.com ("Nope Francis" 2016).
15. See, for instance, Grice (2017), BBC News (2017), Farand (2017), and Roxborough (2017) for discussions of how fake news has affected recent elections in Europe and the 2016 Brexit vote in the UK.
16. Of course, as we discuss in Chapters 3 and 4, fake news, in the form of politically motivated falsehoods, has been with us, in different forms, for a long time. Indeed, it arguably played an essential role in the founding of the United States, as evidenced by the role of political pamphlets and character assassinations in the late eighteenth century (Wood 1993; Chernow 2005). It has also played an important role in the preservation of slavery, colonialism, and systemic oppression of various sorts in democratic societies. (See, for instance, ch. 4 of I. B. Wells [2014]) for a late-nineteenth-century take.) So the issue is not that today's fake news is, ipso facto, a novel problem for democracy. But we do think that (1) fake news, whether new or not, poses a problem for democracy; and that (2) the rise of new media in the twentieth century, and the internet today, has accelerated and expanded the spread of fake news in ways that are different in kind from anything that came before. We are grateful to Liam K. Bright for pushing us on these points.
17. We say more about what we mean by "true" and "false" in Chapter 1—though we make no effort to develop an "account" of truth (or meaning, or belief, etc.). But for the cognoscenti dissatisfied with the discussion in Chapter 1: As far as the metaphysics of "truth" goes, we adopt a broadly deflationary attitude in the spirit of what is sometimes called "disquotationalism." See Ramsey (1927), Field (1986), Maddy (2007, ch. II.2), Burgess and Burgess (2011), and Stoljar and Damnjanovic (1997) for discussions and defenses. But our understanding of truth also has a strong dose of pragmatism: we understand "true beliefs" to be beliefs that generally successfully guide action, and more important, we understand "false beliefs" to be ones that generally fail to reliably guide action. (See, e.g., Ramsey 1931; Skyrms 1984.) But we do not follow traditional pragmatists in holding that truth is somehow defined as or con-

stituted by what one would believe at the "end of inquiry" (Peirce 1878; Misak 2004); our view is that inquiry may well lead us astray, even though there are facts about the ways in which the world is arranged, and the evidence we gather is generally responsive to those facts. This particular mix of disquotationalism and pragmatism strikes us as distinctively Ramseyan.

18. A. F. Smith (1994) gives a detailed history of the tomato's use. He also documents another bizarre incident of false belief. During the mid-1800s, the spread of the tomato worm through the Northeast United States was accompanied by fears that the insect was deadly. Various claims, such as that the bite of a tomato worm could cause death, were widely shared. The *Syracuse Standard* published the account of one Dr. Fuller who claimed that the tomato worm was "poisonous as a rattlesnake" and could throw spittle several feet. "This spittle striking the skin, the parts commence at once to swell, and in a few hours death ends the agonies of the patient" (ibid., 58). This terrifying beast was, in fact, the harmless larva of the hawk moth.
19. For explorations of this perspective, see, for instance, Tversky and Kahneman (1974), Kahneman (2011), and Ariely (2008). See also Festinger (1962).
20. See NSF (2014).
21. Philosopher of science Axel Gelfert (2014) calls this a testimonial conundrum. As he points out, "belief expansion is an epistemically 'risky' move, and for the risk to be worth it, the promise of obtaining new knowledge, from the agent's point of view, must outweigh the danger of acquiring a falsehood" (43). Zollman (2015) uses a simple mathematical model to show how different strategies for acquiring beliefs from the testimony of others can be successful given different goals, such as maximizing one's number of true beliefs, minimizing the number of false ones, or getting the best possible ratio of true to false beliefs.
22. Again, though, see the caveats of note 16.
23. The *Los Angeles Times* has reported on FBI investigations of RT media (Cloud, Wilkinson, and Tanfani 2017), and the EU has identified RT as a source of state-sponsored disinformation (Bentzen 2017).
24. The work of Oreskes and Conway builds on somewhat earlier, equally important work by David Michaels (Michaels and Monforton 2005; Michaels 2008); see also Davis (2002).
25. Here we follow an important recent tradition in philosophy of science and so-called social epistemology, which studies how social factors influence knowledge and belief (Fuller 1988; A. I. Goldman 1986, 1999; Longino 1990, 2002; Gilbert 1992). For an overview, see Goldman and Blanchard (2001).
26. Recall note 19.
27. This has been argued extensively in the field of social epistemology by, for example, Alvin Goldman (1999) and feminist philosophers of science such as Okruhlik (1994) and Longino (1990).

28. This has been called the "independence thesis"—that rational individuals can form irrational groups, and, conversely, irrational individuals can form rational groups (Mayo-Wilson, Zollman, and Danks 2011).
29. For a detailed account of worries about water fluoridation, including discussion of the John Birch Society, see Freeze and Lehr (2009).
30. Pariser (2011) writes extensively about what he calls "filter bubbles"—where individuals prune their social media connections to filter out points of view they do not agree with.
31. A report on National Public Radio (NPR Staff 2016), for example, worried about echo chambers in the lead-up to the 2016 presidential election.
32. This point is also made very nicely by Oreskes and Conway (2010); we return to it at length in Chapter 4.
33. See Schiffrin (2017).

ONE
What Is Truth?

1. This history of the discovery of the ozone hole (Farman, Gardiner, and Shanklin 1985), and its connection to background debates about CFCs, is indebted to Oreskes and Conway (2010, ch. 4). See also Shanklin (2010), one of the coauthors of the original paper, for a discussion of the discovery.
2. For the early history of ozone science, see Rubin (2001, 2002, 2003, 2004).
3. This was despite the fact that the BAS scientists had written to the head of the Satellite Ozone Analysis Center at the Lawrence Livermore National Laboratory in Northern California to inquire about the anomaly. (The BAS team never heard back.)
4. See, for instance, Crutzen (1970), Johnston (1971), McElroy and McConnell (1971), McElroy et al. (1974), Wofsy and McElroy (1974), Stolarski and Cicerone (1974), and Lovelock (1974).
5. Molina and Rowland (1974) first introduced the idea that CFCs could deplete ozone; see also Cicerone, Stolarski, and Walters (1974). Rowland (1989) provides an overview of the history and evidence that CFCs deplete ozone fifteen years after he first proposed as much.
6. According to Google Scholar, Molina and Rowland (1974) was cited 517 times between 1974 and 1985, when the BAS study was released. This surely underestimates the number of studies building on their work and that of Crutzen and others.
7. See Oreskes and Conway (2010, 118–119).
8. Sunstein (2007) discusses the Montreal Protocol at length and compares its success to the failure of the Kyoto Protocol on climate change.
9. See John 18:38.

10. The Rove quotation first appeared, attributed to an anonymous "senior advisor to Bush," in Suskind (2004); since then, it has been widely attributed to Rove (Danner 2007). Kellyanne Conway's reference to alternative facts occurred during an interview with *Meet the Press* on January 22, 2017.
11. See, for instance, McMaster (2011).
12. Again, for a sketch of the views on truth in the background here, see note 17 in the Introduction.
13. McCurdy 1975.
14. Less judicious on the part of industry were the vicious attacks mounted against Sherwood Rowland for his continued research and advocacy on the basis of his findings. See, for instance, Jones (1988).
15. This is reported, for example, by the *New York Times* (Glaberson 1988) and described in Peterson (1999, 246). Curiously, DuPont changed its stance within three weeks of this letter, announcing that the company would suspend all production of CFCs. What ultimately persuaded DuPont? It seems that market forces did most of the work: as awareness of the ozone hole increased, consumers stopped purchasing products with CFCs. By 1988, it became clear that the benefits of continuing to manufacture CFCs were plummeting, even as the costs of defending them increased (Maxwell and Briscoe 1997). Others, however, continued to argue that the case against CFCs was not so clear even into the 1990s—including former Democratic governor of Washington Dixy Lee Ray, in her book *Trashing the Planet* (Ray and Guzzo 1990).
16. Hume presented this work in the *Treatise on Human Nature*, Book 1.iii.6 (1738) and *An Enquiry Concerning Human Understanding*, Book IV (1748).
17. Hume worked in a tradition known as empiricism, which was influential in the British Isles during the seventeenth and eighteenth centuries. (Other well-known philosophers in the empiricist tradition include Francis Bacon, John Locke, George Berkeley, and even, arguably, Isaac Newton.) The empiricists were moved by the basic principle that all of our knowledge, insofar as we have any, must be derived from experience. This idea may sound appealing, and even broadly scientific, but it leads to surprising and unfortunate places, in part because of the Problem of Induction.
18. See, for instance, Weatherall (2016).
19. See Laudan (1981) for the classical version of the argument. Stanford (2001, 2010) connects this to the problem of unconceived alternatives—that throughout the history of science our best theories appeared best because we were, as of yet, unable to conceive of the better alternatives that would come along. These arguments are part of a long-standing debate over "realism" as regards scientific theories. See, for instance, Godfrey-Smith (2009) for a discussion of the issues.
20. This is from *An Enquiry Concerning Human Understanding*, Section X.1 (Hume 1748).

21. See, for instance, Skyrms (1984) for a view very similar to the one we intend to endorse here.
22. For introductions to Bayes' rule, see, for instance, Skyrms (1986) and Hacking (2001); Earman (1992) offers a more sophisticated (and critical) take. The so-called Dutch book argument for the unique rationality of Bayes' rule can be traced back to the work of Ramsey (1931). Hajek (2008) lays out the argument in detail. For an engaging account for nonacademics, see McGrayne (2011); see also Silver (2012).
23. The quotations in this paragraph, and the description of the general character of the criticism of Rowland, come from Jones (1988).
24. A recent accessible overview of the literature on values in science is in Elliott (2017).
25. Bird (2014) provides a detailed biography of Thomas Kuhn. Kuhn's work is an early and highly influential example of the tradition we are considering, but we do not mean to suggest that it does not build on ideas present in even earlier work. In particular, there was a prior tradition in the sociology of science, most famously associated with Robert K. Merton (1942, 1973). Merton, however, unlike the later sociologists of science, does not seem to have taken social factors to matter to the *content* of scientific ideas.
26. Unless we are very sophisticated, and then we see a body traveling along its natural space-time geodesic, suddenly stopped by another body. See Wald (1984) or, for a gentler introduction, Weatherall (2016).
27. Kuhn's key example here was the transition from affinity theory in chemistry to atomism in the early nineteenth century. He claimed that atomism, as developed by John Dalton, made specific testable predictions, and when chemists following Dalton tested those predictions, they found strong agreement with Dalton's theory. But Kuhn observed that very similar experiments had been performed in the late eighteenth century, before Dalton, and the results had been flatly inconsistent with atomism. Kuhn's conclusion was that the change in paradigm changed the experimental results.
28. It was never entirely clear whether Kuhn himself was willing to accept the most extreme form of his own view, but others who were influenced by him certainly did so.
29. Popper (1959) is a good example. See Godfrey-Smith (2009) or Barker and Kitcher (2013) for discussions of philosophy of science before and after Kuhn.
30. In fact, more recent philosophers and historians have argued that the very notion of "scientific objectivity" has changed over time (Daston and Galison 2007).
31. See Cowan (1972) and MacKenzie (1999).
32. See Foucault (2012). Foucault's book was originally published in 1963 and seems to have developed in parallel to Kuhn's work; Foucault's work has also

been profoundly influential in its own right, and we do not mean to suggest that Foucault was merely following Kuhn's lead.

33. See, for example, Dennis (1995).
34. See, for example, Douglas (2009).
35. Kimble 1978; Wang 1992.
36. See Steingrímsson (1998) for a firsthand account of the eruption; various more recent analyses of the eruption and its atmospheric consequences have been done, drawing on a range of sources (Thordarson and Self 1993; Thordarson 2003; Stevenson et al. 2003; Trigo, Vaquero, and Stothers 2010; Schmidt et al. 2012). A nice narrative account is given by *The Economist* (2007).
37. Quoted in *The Economist* (2007).
38. These estimates of the amounts of gas released come from Thordarson et al. (1996). Clouds of ash and noxious gases spread throughout the world, leading to a famine in Egypt that wiped out more than 16 percent of the population (Oman et al. 2006), and in the words of Benjamin Franklin, "a constant fog over all of Europe, and a great part of North America" that led to one of the coldest, snowiest winters on record (still) in New England and the Mid-Atlantic (Franklin 1785, 359).
39. Reed (2014) provides a detailed biography of Angus, including a discussion of his role in early environmental regulation.
40. For a discussion of these sorts of effects, see Likens, Bormann, and Johnson (1972) and Winkler (1976, 2013).
41. These findings were reported in Likens and Bormann (1974); see also Likens, Bormann, and Johnson (1972) and Cogbill and Likens (1974). Likens (1999) provides an overview of the research done on this subject, particularly at the Hubbard Experimental Forest. Oreskes and Conway (2010, ch. 3) provide an in-depth history of the acid rain discovery and ensuing controversy; we rely on them for a number of historical details.
42. These two reports are described in Oreskes and Conway (2010).
43. See *Wall Street Journal* (1982).
44. The claim that the Nierenberg report was tampered with is carefully and extensively documented by Oreskes and Conway (2010, ch. 3).
45. These working groups had concluded, just as every other major scientific body had, that acid rain produced by human activity was causing serious damage. Their work was reviewed in Canada by the Royal Society, and under ordinary circumstances, it would have been reviewed in the United States by a National Academy of Sciences panel. Instead, the Reagan Administration decided to produce its own panel of experts, bypassing the National Academy.
46. See Oreskes and Conway (2010, 86).
47. We searched Google Scholar for articles with the author name "SF Singer" (which is how Singer published) published before 1983, and with keywords

"acid rain" or "acid precipitation"; we also searched for "rain" and "precipitation" and did a manual check for articles that appear to address acid rain. The closest we came was an article on pollution in general (S. F. Singer 1970); in 1984, after already serving on the panel, he wrote a position paper on acid rain in a journal on policy, not science (S. F. Singer 1984).

48. The Rahn quotation comes from M. Sun (1984).
49. See, for example, Franklin (1984).
50. For a discussion of the legislative effort to update the Clean Air Act during 1983 and 1984, see Wilcher (1986).
51. See F. Singer (1996); also quoted in Oreskes and Conway (2010, 133).
52. See, for instance, Ross (1996), for an overview of the issues at stake in the science wars, from the science studies perspective. (This book contains the essays published in the special volume of *Social Text* along with the Sokal hoax paper, as described in note 53.) For a taste of the other side (other than Gross and Levitt [1997], or Sokal and Bricmont [1999]), see, for instance, Newton (1997). Kitcher (2001) offers a compelling middle way that has influenced a great deal of subsequent work in philosophy of science.
53. Perhaps the most famous episode of the whole period was the "Sokal affair," in which Alan D. Sokal, a physicist at New York University, submitted a paper titled "Transgressing the Boundaries: Toward a Transformative Hermeneutics of Quantum Gravity" to an academic journal called *Social Text* (1996). Most of the contributions to the volume in which Sokal's article appeared were by academics working in science studies; Sokal's paper stood out as having come from an actual physicist. It was accepted and published—at which point Sokal revealed that it was intentionally gibberish, written as a parody of the whole genre. He and many others took the paper's acceptance for publication as incontrovertible evidence that the entire field lacked intellectual rigor. (The other papers published with Sokal's were collected in Ross [1996]. One can judge for oneself whether they are all intellectually bankrupt.) Sokal went on to coauthor a book with French physicist Jean Bricmont, published in the US as *Fashionable Nonsense* (1999), which criticized various famous academics as charlatans. We are generally sympathetic with Sokal and Bricmont—and even Levitt and Gross—insofar as they claim that some "postmodernist" writers fetishize obscurantist writing and sloppy thinking. On the other hand, those who study the history, philosophy, and sociology of science have revealed a great deal about scientific thought. Much of what passed for the science wars was little more than a political stunt, orchestrated to make those who sought to break down historic barriers to entry in science for women and members of underrepresented groups look as if they were opposed to science itself. Though he does not put the point in this way, our perspective is strongly influenced by Kitcher (2001).
54. Gross and Levitt (1997, 9) claimed that they were not "stalking horses of

social conservatism," and they attempted to draw a distinction between the "academic left"—a term that was supposed to refer specifically to those whose "doctrinal idiosyncrasies sustain the misreadings of sciences"—and a more general understanding of the political left. But there was nonetheless a clear political subtext to their work.

55. See Lewontin (1998).
56. For example, see Rifkin (1980).
57. Shrader-Frechette (2014) has emphasized this point, that philosophers of science in particular are well-positioned to help expose science built on misguided assumptions.
58. Again, see note 17 in the Introduction. This is where we close out the promissory note.

TWO
Polarization and Conformity

1. See Clarkson (1997).
2. See Zhao, Zhu, and Sui (2006).
3. See Keynes (2008), who writes at length about the madness at the end of Newton's life.
4. Stokes and Giang (2017) provide a list of current international mercury regulations.
5. Most of this history is drawn from Hightower (2011).
6. See Rice, Schoeny, and Mahaffey (2003).
7. This is according to Hightower (2011).
8. ABC News (2006) reports on the episode.
9. See, for example, chapter 3 of Weatherall (2016).
10. This now famous anecdote is relayed by James Watson (2011).
11. Among these were data on electron bond angles and images of the molecule developed by Rosalind Franklin at King's College. Again, the lone (or in this case pair) genius story does not fit the bill. See Olby (1974) for a history of the episode.
12. For readers unfamiliar with Tinkertoys, they are sets of interlocking building parts designed for children to play with.
13. Of course, the practice of modeling is wildly heterogeneous (see, e.g., Downes 1992; O'Connor and Weatherall 2016). We do not mean to imply that all scientific models involve simplified systems intended to reveal truths. Predictive models of weather patterns, for example, are extremely complex and not typically intended to provide insight into how the system works. Weisberg (2012) provides an in-depth discussion of many ways models are used in science.

14. We choose one framework to simplify things for the reader and because we prefer to focus on what we take to be our original contributions. We use this particular framework because it fits well as a representation of science, but we do not mean to suggest that this is the only option. To give some salient examples: There is a robust and long-standing literature on the diffusion of innovations or ideas in connected social networks, and on similarities between this process and processes of contagion. Some of these models make some variant on the assumption that individuals simply adopt ideas or behaviors they are exposed to (see, e.g., Abrahamson and Rosenkopf 1997; Allen 1982; Rogers 2010; and Kleinberg 2007). Others assume that individuals consider the distribution of behaviors and beliefs among their network neighbors in deciding whether to adopt (as in Young 2006a, 2001; Montanari and Saberi 2010). Yet others focus on the particular assumption that individuals adopt ideas or behaviors that have passed some threshold of prevalence (Granovetter 1978; Deffuant, Huet, and Amblard 2005; Weisbuch et al. 2002). Some authors in the marketing literature have studied how interventions might improve the spread of innovations (Choi, Kim, and Lee 2010). With some tweaking these models could be modified to address the spread of novel scientific ideas instead. Some results from this literature, such as that the speed of contagion is deeply influenced by network structure, are very relevant. Other relevant models focus on "opinion dynamics"—the spread of variant opinions on a network under different assumptions about agents and about the way opinions spread. These opinions sometimes matter to player payoff (Ellison and Fudenberg 1995), but typically do not. For some examples, see Arifovic, Eaton, and Walker (2015); Jalili (2013, 2015); Holyst, Kacperski, and Schweitzer (2001); Lu, Sun, and Liu (2017); and Golub and Jackson (2007). In addition, some authors, especially recently, have begun to explicitly model the spread of false beliefs through social media and social networks (as in Ioannidis and Varsakelis 2017).
15. See Zollman (2007) and Bala and Goyal (1998). Other articles in this tradition in philosophy include Zollman (2010a, b); Mayo-Wilson, Zollman, and Danks (2011); Zollman (2013); Kummerfeld and Zollman (2015); Holman and Bruner (2015); Borg et al. (2017); Rosenstock, Bruner, and O'Connor (2016); and Holman and Bruner (2017).
16. Bruner and O'Connor (2017) do use models to address the role of power in the emergence of scientific norms. Other philosophers of science use models to address diverse aspects of science, from the gender publication gap (Bright 2017) to citation rates (Heesen 2017a).
17. In fact, this sort of model is sometimes called a "two-armed bandit problem" for this very reason, because "one-armed bandit" is another name for the casino game.

18. Real human networks are often "small worlds," which tend to have interconnected cliques and short path lengths between any two nodes. Granovetter (1973) conceived of human networks as consisting of tight-knit cliques connected by bridges. Watts and Strogatz (1998) developed their famous "small world" networks to try to capture realistic aspects of human networks. Onnela et al. (2007) observe such structures among cell phone users. Newman (2001) finds the same for scientific collaboration networks.
19. When credences exactly equal .5, we also have the agent choose action A, but this essentially never happens.
20. Zollman (2007, 2010b) first made the connection between these models and doctors' choices between treatments.
21. The numbers here assume that action B pays off with probability .6, and action A pays off with probability .5. The agents know that action A pays off with this fixed probability, and they are unsure about whether action B pays off with probability .6 or .4 (but they know that these are the only two possibilities).
22. One modeling approach in philosophy of science very similar to what we employ is the network epistemology framework Laputa (Angere n.d.). The main difference is that the Laputa framework involves agents who share opinions rather than evidence, which we take to be a less accurate representation of scientific sharing but a more accurate representation of everyday belief transmission. This framework has been used to investigate belief polarization (Olsson 2013), confidence in one's epistemic abilities (Vallinder and Olsson 2014), norms of assertion (Olsson and Vallinder 2013), and the influence of network structure on belief accuracy (Jönsson, Hahn, and Olsson 2015; Angere and Olsson 2017). These results and others are presented in Angere, Masterton, and Olsson (n.d.).
23. For more on the discovery of *Helicobacter pylori*, see Warren and Marshall (1983) and Marshall and Warren (1984). For parts of this history, we depend on Zollman (2010b), who uses this example to illustrate epistemic network models.
24. See a detailed history in Kidd and Modlin (1998).
25. See Palmer (1954).
26. This is reported autobiographically by Barry Marshall (2002).
27. See, for example, proofs from Bala and Goyal (1998) and simulations from Zollman (2007), who also shows (2010b) that under extreme conditions, scientists can maintain different beliefs for long periods of time in such models.
28. Rosenstock, Bruner, and O'Connor (2016) show that Zollman's research on connectivity and consensus is sensitive to parameter values and that, in particular, in communities where scientists gather large data sets and data is more

reliable, it is better to always communicate this data. Generally, their work shows that the harder the problem, the more potential social structure has to influence the outcome.

29. This result is due to Zollman (2007, 2010b), as is the connection to the case study on *H. pylori* and stomach ulcers.
30. Another way to maintain this diversity of beliefs for a long enough time is if researchers start off very convinced in different theories. This translates to having very high credences in different beliefs in this sort of model, as Zollman (2010b) shows. Other philosophers of science have considered how the credit economy—the fact that scientists are motivated, to some degree, by the desire to obtain academic credit—can incentivize scientists to test different theories, even if they all think one is most promising. Kitcher (1990) and Strevens (2003) both propose influential models showing how credit can improve the division of cognitive labor in a scientific community.
31. Zollman (2010b) first connected this case to the sort of model he develops for the detrimental effect of oversharing evidence.
32. Relatedly, Kummerfeld and Zollman (2015) show how scientists who are more adventuresome, in that they test theories they do not necessarily think are the best, do not fall prey to the Zollman effect. Because they naturally introduce a diversity of behavior into their network, the group does not need to rely on limited communication to ensure this sort of diversity.
33. Murray gives her account of her experiences in *The Widening Circle: A Lyme Disease Pioneer Tells Her Story* (1996).
34. Some of this history is drawn from David Grann's *New York Times* article "Stalking Dr. Steere over Lyme Disease" (2001). For further insight, see Specter (2013).
35. Steere and coauthors described their early findings in "An Epidemic of Oligoarticular Arthritis in Children and Adults in Three Connecticut Communities" (1977).
36. See Burgdorfer et al. (1982).
37. See Auwaerter, Aucott, and Dumler (2004).
38. See Halperin (2008).
39. See Bratton et al. (2008) and Fallon and Nields (1994).
40. See CDC (2015) for guidelines for Lyme disease.
41. See Feder et al. (2007).
42. See Feder et al. (2007) and Marques (2008) for a discussion of these studies.
43. Some of this history is drawn from the *New Yorker* article "The Lyme Wars" (Specter 2013).
44. Straubinger 2000; Straubinger et al. 2000; Embers et al. 2012; Bockenstedt et al. 2002.
45. See Embers et al. (2012) and Bockenstedt et al. (2002). More recently see Embers et al. (2017) and Crossland, Alvarez, and Embers (2017). Marques et

al. (2014) retrieved Lyme DNA from a sterile tick that had fed on a human previously treated with antibiotics for Lyme. This does not definitively prove that the Lyme spirochetes detected were alive, however. Stricker (2007) provides an overview of research to that date.

46. For example, see the documentary *Under Our Skin* (Wilson 2008).
47. The documentary *Under Our Skin* reports on the disciplining of Dr. Jones and other Lyme-literate physicians (Wilson 2008).
48. See Shear, Goldman, and Cochrane (2017).
49. See Zadronsky et al. (2017).
50. See Bromwich and Blinder (2017).
51. See Koerner and Lewis (2017).
52. Feminist philosophers of science and feminist epistemologists have done more than perhaps any other group to highlight the value-ladenness of science. For examples, see Longino (1990); Haraway (1989); and Okruhlik (1994). See also Kitcher (2001, 2011).
53. We should not overstate this point. It may well be that some establishment physicians have been swayed by insurance funding. It may also be that some Lyme-literate physicians are in it for the money. The point is that most of those involved in the Lyme wars seem primarily motivated by a desire to cure patients.
54. Essentially all extant models that capture a community fracturing into groups that hold stable, polarized opinions adopt some version of the assumption that social influence between individuals is mediated by the similarity of their beliefs. Most of these models look at polarization of opinions where all the options are essentially equally useful. For example, Hegselmann and Krause (2002) introduce a widely employed model in which individuals adopt opinions that are numbers between 0 and 100. For some set of opinions near their own (say five units above and below), every round, individuals will change their opinions to the average of this set. Over time, we see movement toward consensus. If the sets are big enough, everyone always ends up with the same opinion. But if the sets are small, subgroups emerge with different, stable beliefs. In other words, by adding the fact that individuals are influenced only by those with opinions like their own, the models end up with polarization. (See also Deffuant et al. 2002; and Deffuant 2006.) Although the details vary widely, models from R. Axelrod (1997), Macy et al. (2003), and Baldassarri and Bearman (2007), where polarization arises, all include an assumption of this sort. There are two models that, like ours, show polarization in cases where one belief is better than another. These are from D. J. Singer et al. (n.d.) and Olsson (2013). In the Singer et al. model, actors share evidence, but this evidence does not map onto the sort of data gathered in a scientific experiment, as in our model. The Olsson model is quite similar to ours, as it uses a network structure and Bayesian updating, but the actors state their

opinions rather than sharing data. In O'Connor and Weatherall (2017) we focus on explaining how scientists, like those investigating Lyme disease, could polarize despite sharing values and data. See also models from Galam and Moscovici (1991); Galam (2010, 2011); Nowak, Szamrei, and Latané (1990); Mäs and Flache (2013); and La Rocca, Braunstein, and Vazquez (2014).

55. Jeffrey laid out this updating rule in *The Logic of Decision* (1990).
56. Some psychologists and others have argued for the realism of this sort of reverse updating—sometimes dubbed the "backfire effect" or "boomerang effect"—especially in cases where individuals are politically polarized. (See, e.g., Nyhan and Reifler 2010.) Other studies have failed to replicate this effect, as in Cameron et al. (2013). See Engber (2018) for a popularized discussion of this research.
57. See O'Connor and Weatherall (2017). "Polarization" could mean a lot of things (Bramson et al. 2016). In our work, we focus on a broad set of outcomes where a community breaks into two groups, one whose members have very high beliefs in the correct theory, and the other whose members believe in the incorrect theory but are not influenced by the rest of the community.
58. Jern, Chang, and Kemp (2014) provide an overview of this literature while showing how, contra claims in psychology, those with different background assumptions can rationally polarize when presented with the same evidence.
59. See Taber, Cann, and Kucsova (2009), though Kuhn and Lao (1996) argue that this sort of outcome, while real, is less prevalent than some psychologists would claim. See also Engber (2018).
60. Semmelweis documented this history in his 1861 book *The Etiology, Concept, and Prophylaxis of Childbed Fever* (Semmelweis 1983).
61. For more on this, see Carter (2017).
62. See Wallace (2017) and Ford (2017).
63. See, for example, Fandos (2017).
64. See Schaffner and Luks (2017) and Levinovitz (2017).
65. See Asch (1951).
66. See Bond and Smith (1996) and Padalia (2014).
67. See Condorcet (1785) for his original work.
68. Likewise, Zollman (2010a) shows how conformity bias can improve the judgments of a group under some situations. In his models, members of a networked group start with private information, make public statements based on this information, and then, in every following round, make public statements that conform to the majority of their neighbors. Sometimes these groups will be quite good at reaching true beliefs since their conformity aggregates the original, partially dependable private information.
69. James Surowiecki (2005) presents this example in his book *The Wisdom of Crowds*, which also discusses crowd-based knowledge in much greater depth and discusses information cascades.

70. See Bikhchandani, Hirshleifer, and Welch (1992) for an introduction to the idea of information cascades, and also for some good examples of conformity resulting from information cascades. See Easley and Kleinberg (2010) for more on the topic. Models of persuasion bias also explore how sharing interdependent opinions can lead to false beliefs (DeMarzo, Vayanos, and Zwiebel 2003).
71. Asch 1951, 227.
72. To be clear, conformity bias might not be the only thing going on in these cases. For example, physicians may have been horrified at the idea that they were directly responsible for patient deaths and thus were unwilling to give merit to Semmelweis. Trump supporters may have been motivated to spurn the mainstream media criticizing Trump and Spicer.
73. We draw most of this discussion from the work presented in Weatherall and O'Connor (2018). We also draw on the work of Mohseni and Williams (n.d.) and Schneider (n.d.), as noted.
74. Many of the extant models exploring conformity and behavior have a flavor similar to models of information cascades (discussed above), as in Banerjee (1992); Buechel, Hellmann, and Klössner (2015); and Walden and Browne (2009). Some models of conformity are designed to address the phenomenon of pluralistic ignorance, where many individuals privately doubt a publicly supported belief or norm (Huang, Tzou, and Sun 2011). Duggins (2017) considers agents with varying psychological tendencies, including conformity but also intolerance, susceptibility, and the drive to be distinct, to show how much diversity of opinion may be sustained in such a model.
75. Schneider (n.d.) examines a model of this sort intended to represent scientific communities. Agents can choose one of two beliefs and base their choice on a desire to conform with neighbors. His interpretation of this model is that scientists garner real payoffs from conforming, since there are upsides to coordinating theory choice in science. He explores the possibility that one belief might be better in that conforming with others yields more benefits. As he shows, even when one belief is better, though, it might not be adopted by all the agents if they cluster into social groups. In addition, previous authors have used these sorts of models to investigate the dynamics of innovation for different network structures. For examples, see Young (2006a, 2006b, 2011).
76. See Schneider (n.d.).
77. In fact, one can choose the network such that *any* fraction you like, no matter how close to 100 percent, will continue to perform the worse action, even though everyone holds the true belief. This is so even if both performing the true action and conforming matter to everyone, as long as conformity matters enough.
78. See Mohseni and Williams (n.d).

79. For this reason, the spread of Vegetable Lamb–like beliefs might, in fact, be more fruitfully represented by a branch of modeling that looks at the spread of rumors and misinformation. Daley and Kendal (1965), for example, in a very early model, look at the random diffusion of rumors under the assumption that individuals will spread or stifle a rumor on the basis of whether the rumor seems out of date. Dabke and Arroyo (2016) use this framework to explicitly model the spread of information on social media. Others consider differences in the personalities or power of agents as a factor influencing misinformation spread, as in Acemoglue, Ozdaglar, and ParandehGheibi (2010). See also Zhao et al. (2011).
80. See Baron, Vandello, and Brunsman (1996). Remarkably, though, in a variant on the Asch experiment where it is more difficult to choose the right line, monetary incentives increase conformity. This undergirds the claim that some conformity is a rational response to uncertainty, while other conformity arises as a direct desire to do the same thing as others.
81. In the case of irradiation, Conley (1992) documents widely held fears about irradiation at the time, despite the safety of irradiated foods (Joint FAO/IAEA/WHO Study Group 1999). Funk and Rainie (2015) found a wide gap between scientists and the American public in opinions of the safety of genetically modified foods. Massive protest efforts have been aimed at curbing the use of genetically modified organisms (GMOs) (see, e.g., Kuntz 2012), despite little evidence that they pose hazards (Nicolia et al. 2014). While many consumers believe that organic food contains more nutrients than its counterparts, the evidence for this is mixed (Smith-Spangler et al. 2012; Hunter et al. 2011).
82. See, for example, Zucker et al. (2017).

THREE
The Evangelization of Peoples

1. See Norr (1952). The history presented here draws on the account given by Oreskes and Conway (2010, ch. 1), though note that the chronology we describe differs slightly from theirs, as they report that the *Reader's Digest* article was published in the wake of Wynder, Graham, and Croninger (1953), whereas in fact it appeared almost a year earlier. For further background on the fight over the regulation of tobacco products, see Kluger (1997), Brandt (2009), and especially Koop (1998). Many documents related to the tobacco industry and its efforts in connection with the regulation of tobacco products are available through the Tobacco Control Archives at UC San Francisco (UCSF Library n.d.).
2. For a history of *Reader's Digest* through this period, see Heidenry (1995).

3. See Wynder, Graham, and Croninger (1953).
4. *Time* 1953.
5. See *New York Times* (1953a).
6. See *New York Times* (1953b).
7. Dorn 1954, 7.
8. Although we are following the account by Oreskes and Conway (2010), they were not the first to identify and characterize various aspects of the Tobacco Strategy; for instance, David Michaels called this strategy "manufacturing uncertainty" in his work on industry and public health (Michaels and Monforton 2005; Michaels 2008). See McGarity and Wagner (2008) for an excellent analysis of the types of strategies that industrial propagandists can take in general.
9. The memo can be found in the Tobacco Control Archives (UCSF Library n.d.). This same memo goes on to acknowledge that "doubt is also the limit of our 'product.' Unfortunately, we cannot take a position directly opposing the anti-cigarette forces and say that cigarettes are a contributor to good health. No information that we have supports such a claim."
10. The "Frank Statement" can be found in the Tobacco Control Archives (UCSF Library n.d., doc ID: zkph0129).
11. Tobacco firms themselves acknowledged this during the 1950s. For instance, one memo produced by Liggett & Myers, a tobacco firm that did not participate in the TIRC, stated in 1958 that "the constantly reiterated 'not proven' statements in the face of mounting contrary evidence has [*sic*] thoroughly discredited the TIRC, and the SAB [Scientific Advisory Board] of TIRC is supporting almost without exception projects that are not related directly to smoking and lung cancer" (quoted in *State of Minnesota and Blue Cross Blue Shield of Minnesota v. Philip Morris, Inc.* 1998).
12. Oreskes and Conway (2010, 17), for instance, describe the TIRC's support of Wilhelm Hueper, a National Cancer Institute researcher who focused on asbestos and who often testified in court that asbestos, and not tobacco, was the cause of particular patients' cancers.
13. See, for instance, the Teaque memo from 1953 quoted in Cummings, Morley, and Hyland (2002).
14. Data on total sales (or actually, total production, which is an imperfect estimator) comes from Burns et al. (1997). Note that although total production increased through the early 1980s, per capita consumption increased only until the early 1960s, stayed steady until the late 1960s, and then has dropped ever since. This does not mean that smoking did not increase—rather, it means that smoking increased only proportionately with population increases.
15. See, for instance, Dunn (1979) for an overview of the religious wars in Europe during this period.
16. See Marx and Engels (1906, 51).

17. A history of the CPI's efforts is given by A. Axelrod (2009). Bernays (1942) referred to these activities as psychological warfare in a later article.
18. For the editorial on the Committee on Public Misinformation, see *New York Times* (1917). It was Edward Bernays (1942) who described the CPI as part of America's "psychological warfare" effort; Walter Lippmann also used the phrase (Kennedy 1980, 91).
19. See Kearns, Schmidt, and Glantz (2016).
20. See Grim (2010).
21. For an engaging overview of the various strategies taken by industry to influence science since the invention of the Tobacco Strategy, and numerous case studies, see McGarity and Wagner (2008).
22. Bernays 1928, 9.
23. See Advisory Committee to the Surgeon General of the Public Health Service (1964).
24. This timeline is taken from the US Department of Health and Human Services website (2012).
25. See Christakis and Fowler (2008).
26. In a previous study Christakis and Fowler (2007) found similar effects on obesity—that those who lost weight influenced their friends' chances of losing weight. Later (Christakis and Fowler 2009) they discuss these and other findings related to the way people influence each other's behavior through social networks.
27. Notice that the conformity models in the previous chapter may be relevant to understanding social pressures related to smoking and smoking cessation.
28. This model is described at length in Weatherall, O'Connor, and Bruner (2018). We emphasize that the work by Holman and Bruner (2015, 2017) on intransigently biased agents and industrial selection that we discuss below, which uses the same modeling framework to address different aspects of industrial propaganda in scientific communities, was done before our work on the Tobacco Strategy and served as inspiration for our own modeling. We lead with our work because, again, we prefer to focus in the present book on what we take to be original and because our principal focus in this chapter is on how ideas from science flow to the public, not on interventions that act on scientific communities themselves (though that is surely also relevant).
29. In addition to the accounts given by Oreskes and Conway (2010), see, for instance, Koop (1998); Cummings, Morley, and Hyland (2002); Michaels and Monforton (2005); Michaels (2008); Brandt (2009); and Smith-Spangler et al. (2012) for discussions of the tobacco industry's strategies.
30. These numbers are quoted in Warner (1991), from a document produced by the Tobacco Institute.
31. There are some similarities here to the modeling work of Holman and Bruner (2015), which we discuss at further length later on. They consider a propa-

gandist who pulls a biased bandit arm—that is, for whom B will appear to be worse than A on average—and who shares these results within a scientific community with the aim of persuading this community to adopt action A. The principal differences are that (1) they consider a biased arm, rather than sharing only some of the results produced, which means their model does not capture the idea that under some circumstances, biased production is prohibitively expensive; and (2) their propagandist shares evidence with other scientists, that is, other agents who are gathering their own evidence, which changes the dynamics because it means that agents influenced by the propagandist can then choose not to go on testing action B, whereas in our model, those choices are made independently of any decisions by the propagandist. Like us, they find that low-powered studies are particularly useful to the propagandist, though for different reasons: in our model it is because of the relative costs of studies and the rates at which spurious results arise, and in theirs it is because their scientists are actively trying to identify propagandists, and this is more difficult when the distributions over possible results are comparatively wide.

32. Philosopher of science Felipe Romero (2016) has used models to show how in cases in which scientists suppress data that are not significant, or that do not show a positive result, communities can end up with incorrect beliefs, even under the assumption that they have the resources to run unlimited studies on a phenomenon.
33. Rosenthal (1979) shows how to fill in the gaps in a large data set that is missing null results; however, Scargle (2000) criticizes these methods. He argues instead for requiring preregistration of all scientific studies before they are carried out, a practice that is already being implemented in some cases and should greatly ameliorate the effects of biased production.
34. The sociologist of science Robert Merton (1973) has pointed out that scientific communities tend to conform to a "communal norm"—that scientific work must be shared. Putting unpublished research papers in the file drawer does conflict with this norm, but, as mentioned, it is still standard scientific practice. For modeling work on communist norms in science, see Strevens (2017) and Heesen (2017b).
35. For an example of this newsletter, see the Tobacco Institute, Inc. (1958).
36. We could not find a book in which this quotation appeared.
37. To give a more recent example of selective sharing, from 1987 to 2004, thirty-eight FDA-funded studies showed the efficacy of antidepressants, and thirty-six studies found no benefit from their use. Whereas all the positive studies were published, only three of the negative studies were (Turner et al. 2008). Elliott and Holman (n.d.) assess both the benefits and detriments of industry-funded science in general, including a discussion of this case.
38. We suppose that each round involves ten draws per scientist and that the

agents are trying to determine whether the probability of action B paying off is either .6 or .4.

39. See, for instance, Button et al. (2013), who argue that neuroscience studies have, on average, very low statistical power and that this reduces the probability that even statistically significant results reflect true effects. (Responses to their article question whether low-powered studies actually lead to false results—but do not seem to deny that neuroscience relies on "small-scale science.") Szucs and Ioannidis (2017) provide further empirical support for the claim that neuroscience and psychology rely on low-powered studies. Worries have been raised in other fields, too, such as ecology (Lemoine et al. 2016) and evolutionary biology (Hersch and Phillips 2004).
40. Philosopher of science Remco Heesen (2015) has argued that scientists are sometimes justified in collecting a very small amount of evidence given the trade-off between the cost of being wrong and the cost of collecting data on any topic. When we zoom out to the community level, though, what makes sense for an individual scientist here may put the wider community at risk.
41. For basics on statistical power, see Cohen (1992).
42. In addition to the references in note 40, Smaldino and McElreath (2016) provide a meta-analysis of statistical power from across behavioral sciences over the past sixty years showing that it has not improved despite repeated calls to do something about it. These authors show how the processes of publication and hiring might contribute to the continued production of underpowered studies.
43. See Open Science Collaboration (2015).
44. See Baker (2016).
45. See Benjamin et al. (2017).
46. Much of this history is drawn from Hightower (2011).
47. See Davidson et al. (1998); Davidson et al. (1995); Davidson, Myers, and Weiss (2004); and Myers et al. (1995).
48. See Grandjean et al. (1998).
49. See Hightower (2011) and Davidson (2003).
50. See Hightower (2011) and JIFSAN (1999).
51. See Hightower (2011, 104).
52. See also Smaldino and McElreath (2016) for another model treating a scientific community as a Darwinian-type population.
53. Matthew 25:29: "For whoever has will be given more, and they will have an abundance. Whoever does not have, even what they have will be taken from them." The influential sociologist of science Robert Merton (1968) was the first to apply this term to academic communities.
54. See Moore (1995)
55. See Morganroth et al. (1978) and Winkle (1978).
56. See Moore (1995).

57. CAST 1989.
58. See Moore (1995).
59. Holman and Bruner (2015) refer to this sort of propagandist as an "intransigently biased agent" and focus, in particular, on ways that scientific communities might avoid the negative effects of such an agent. Also see note 32.
60. Bernays 1928, 76.
61. The biographical material presented here largely follows Nierenberg's obituary for Revelle (Nierenberg 1992).
62. Revelle and Suess 1957.
63. See Arrhenius (1896) and Chamberlin (1899).
64. See "1990 AAAS Annual Meeting" (1989).
65. The story is told in detail by Oreskes and Conway (2010, 93–94); see also the affidavits written by Justin Lancaster (Revelle's research assistant) and Christa Beran (Revelle's secretary) and the deposition of Singer produced as part of Singer's lawsuit against Lancaster (*S. Fred Singer v. Justin Lancaster* 1993). Revelle's daughter wrote an op-ed piece in which she discussed her understanding of her father's views (Hufbauer 1992); as noted in the text, Singer later produced his own account of the whole affair (S. F. Singer 2003). Lancaster's version of events was described in a 2006 statement (Lancaster 2006) and in an interview with Oreskes and Conway.
66. Singer 1990, 1139; Singer, Revelle, and Starr 1991, 28.
67. These assertions are made, for instance, in the affidavits written as part of *S. Fred Singer v. Justin Lancaster* (1993).
68. Easterbrook 1992, 24.
69. See Bailey (2016, 28).
70. A related example comes from the use of "key opinion leaders" by pharmaceutical companies. These are doctors paid to act as spokespeople for particular drugs. (See, e.g., Krimsky 2004; Moynihan 2008; Proctor 2012; Sismondo 2013; and Elliott and Holman 2018). In addition, Holman (2015) draws out the general tactic by industrial interests of convening groups of carefully selected "experts" to bolster a convenient scientific claim.
71. The movie *Merchants of Doubt* (Kenner 2014) portrays Singer blatantly discussing NIPCC strategies.
72. There is a second phenomenon here, related but, we think, essentially different: there are websites devoted to "doxing," that is, revealing contact information about scientists who work on global climate change. In some cases, the claims made on such websites attack the credibility of individual scientists—such as James Hansen, a NASA scientist who was the first to testify to Congress, in 1988, that global warming had already been detected. But in many cases, the goal seems to be pure intimidation. Once contact information is posted, scientists begin receiving emails that contain personal attacks and death threats.

73. This story is told in detail in many places, but the best treatment we know of is that given by Oreskes and Conway (2010, ch. 6). See also Bolin (2008) for a history of the IPCC from the perspective of its first chairman.
74. Seitz 1996.
75. Bramson et al. (2016) call this kind of clustering of beliefs "belief convergence."
76. Similarly, some models consider agents who may pretend to share opinions with neighbors to influence them (Afshar and Asadpour 2010; Fan and Pedrycz 2016).
77. Barrett, Skyrms, and Mohseni (2017) show how communities that learn to listen to those who accurately test the world have better beliefs. See also Douven and Riegler (2010) and Acemoglue, Bimpikis, and Ozdaglar (2014).
78. This history is drawn from Grundy (1999).
79. See Grundy (1999).
80. See Riedel (2005).
81. Grundy reports that when Lady Mary asked Maitland to variolate her young daughter after returning to England, he "at first declined to act. He had his career in mind: for a mere surgeon to perform inoculation in London under the eyes of the College of Physicians was a much graver affair than doing it in faraway Turkey" (1999, 210). He eventually consented to perform the procedure.
82. As we have hinted, position in a network is far from the only relevant factor determining influence on others. Some individuals are more influential than others because of personality, prestige, or power, independent of their network connections. In the case of variolation the social status of Princess Caroline and Lady Mary Montagu was an important factor as well as their centrality. In addition, some individuals are more attuned to social and conformist influences than others, which our models do not account for but which will matter for the spread of beliefs. (See Wagner et al. [2012] for an example.) Along these lines X. Chen et al. (2016) employ a network model to argue that manipulating expert opinion (where experts have special ability to influence others) may be more effective than manipulating those who simply have many connections. And Förster, Mauleon, and Vannetelbosch (2016) consider opinion dynamics under the possibility that agents might pay for the ability to manipulate the beliefs of others.
83. See P. J. Smith, Chu, and Barker (2004) for an overview. Unvaccinated children tend to be white, to come from wealthy families, and to cluster geographically, which, ironically, increases the chances of epidemics as a result of anti-vaxxing attitudes.
84. See X. Chen et al. (2016) and Bloch, Keller, and Park (2015).
85. See Molteni (2017)
86. See Wakefield et al. (1998) for the now-retracted article drawing a spurious

connection between vaccines and autism. The *Washington Post* reported on the actions of antivaccine advocates in Minneapolis in May and August 2017 (L. H. Sun 2017a, 2017b).

87. Marketers have learned to use the same effects to their advantage. Choi, Kim, and Lee (2010), for example, present a model in which they argue that marketing to a clique is often especially successful. Social media marketing, in particular, takes advantage of conformity. Seeing that a friend has "liked" something on Facebook has been found to double the chance that peers also "like" it (Egebark and Ekström 2011). It takes three strangers to generate the same effect. Since "likes" can sell a product, it pays to be savvy about the effects of conformity in social media by targeting close-knit groups of people to see the same ads (Yeung and Wyer 2005).

FOUR
The Social Network

1. See Adam Goldman (2016) for Welch's own description of events; see also Fisher, Cox, and Hermann (2016); Lipton (2016); and Weiner (2016). Robb (2017) provides an in-depth analysis of the whole sequence of events.
2. See Silverman (2016a).
3. There is no doubt that Podesta's account was hacked and that many of the emails released were real; it is unclear whether all of the material released was unaltered (Cheney and Wheaton 2016).
4. See Aisch, Huang, and Kang (2016).
5. See Rosenberg (2017).
6. See Aisch, Huang, and Kang (2016) and Fisher, Cox, and Hermann (2016).
7. Welch is quoted in Adam Goldman (2016).
8. Wood 1993; Chernow 2005.
9. Woolf (2016) covers this story. See also Krauss (1998). This history is presented more fully in W. J. Campbell (2001).
10. More recent investigations have consistently come to contrary conclusions. Some have found that the *Maine* was sunk by an external source, such as a mine, perhaps placed by a foreign adversary; others have found that the wreck is most consistent with an *internal* explosion, likely spontaneously produced in the ship's coal storeroom. See, for example, Wegner (2001).
11. See Keller (1969) for a history of the Spanish-American War.
12. This incident is discussed in Thornton (2000).
13. Collins 2017.
14. See Poe's *Balloon-Hoax* (1844). Scudder (1949) discusses the incident.
15. The alleged circulation was printed on the cover of the *Journal.* See also W. J. Campbell (2001) and Smythe (2003). The population figures and ag-

gregate New York newspaper circulation figures come from the *Twelfth Census of the United States*, Vol. 9 (1902, 1051).

16. See Bhagat et al. (2016).
17. See Greenwood, Perrin, and Duggan (2016).
18. See Fiegerman (2017) and G. Wells (2017); compare with the approximately 245 million American adults. (Of course, not all Twitter users are adults, and so the 30 percent figure is merely approximate.)
19. The term "meme," short for *mimeme,* is ancient Greek for "imitated thing." The term was proposed to describe units of cultural transmission in Dawkins (2006).
20. See Singer-Vine and Silverman (2016).
21. See, for instance, Pennycook and Rand (2017) and Allcott and Gentzkow (2017).
22. Barthel, Mitchell, and Holcomb 2016.
23. For models addressing media influence on public belief, see Candia and Mazzitello (2008); Rodríguez, Castillo-Mussot, and Vázquez (2009); and Pinto, Balenzuela, and Dorso (2016).
24. The details of these results turn out to depend on the statistical features of the problem at hand—for instance, whether there tend to be more ways to get results exceeding a given p value that point in one direction as opposed to the other. This sort of behavior is surely an artifact of the model, and it makes it difficult to extract from these models any concrete claims about the conditions under which journalistic practices are most likely to have deleterious effects. On the other hand, these pathologies actually underscore the point we are trying to make, which is that the sorts of distortions of the total body of evidence that we are considering here can have unpredictable and significant effects. Whatever else is the case, intervening in the ways the journalist does in the model should not be expected to be truth-conducive for consumers of journalism.
25. These results are discussed in more detail in Weatherall, O'Connor, and Bruner (2018).
26. This is a point that Oreskes and Conway (2010) emphasize very compellingly.
27. See Iraq Body Count (2013). The Iraq Body Count project has been criticized on both sides, and other estimates have ranged from the mid-tens of thousands to six hundred thousand or more civilian deaths.
28. *New York Times* 2004.
29. And they have been—see, for instance, the discussion of the consumer price index in Chapter 8 of Weatherall (2013).
30. This story has been discussed at great length by many news sources, not all of them reliable. We are relying in particular on stories in the *Los Angeles Times* (Shalby 2017) and CNN (Cillizza 2017) for the general timeline of events.
31. Rich's DNC job is described in Morton (2016).

32. See kurtchella (2016).
33. See Knight (2003).
34. See Matrisciana (1994).
35. CNN Library (2017) provides a timeline of the hack and subsequent fallout.
36. See Tran (2016).
37. See, for instance, the joint statement by the Department of Homeland Security and the director of National Intelligence released on October 7, 2016 (Department of Homeland Security 2016); see also ("Assessing Russian Activities" 2017).
38. See Ewing (2016).
39. See, for instance, RT coverage for July 28, 2016 (RT American 2016) and August 10, 2016 (RT International 2016).
40. See Stuart (2017).
41. See Folkenflik (2017).
42. Concha 2016.
43. Stein 2016.
44. See Ember and Grynbaum (2017).
45. After the two false stories, CNN not only retracted them but fired three journalists who worked for their investigative reporting team.
46. At least, what we say is generally true for print media; policing "news" on television is far more difficult and often less successful.
47. A detailed study of the role of fake news in setting the journalist agenda is provided by Vargo, Guo, and Amazeen (2017); they find that fake news does not exert "excessive power" in setting the media agenda, but it does seem to have a strong influence on partisan media.
48. See Bump (2017), though reporting on this is ongoing, and so new evidence is constantly coming to light.
49. See CNN Library (2017).
50. Isaac and Wakabayashi 2017; Solon and Siddiqui 2017.
51. This was reported in the *New York Times* (Isaac and Shane 2017).
52. See O'Sullivan and Byers (2017).
53. For examples, see Tacchini et al. (2017); Figueira and Oliveira (2017); and Shao et al. 2017). For older proposals, see Gupta et al. (2013) and C. Chen et al. (2013). The website Fake News Challenge (fakenewschallenge.org) encourages computer scientists to compete to identify the best algorithmic approaches to the fake news problem.
54. This quotation can be found in Strathern (1997, 308). Campbell's law, introduced by psychologist Donald Campbell, has a similar flair. He wrote, "The more any quantitative indicator is used for social decision making, the more subject it will be to corruption pressures and the more apt it will be to distort and corrupt the social processes it is intended to monitor" (D. T. Campbell 1979, 85).

55. See Vann (2003).
56. See Holman (n.d.).
57. Pariser (2011) discusses both the reality of these so-called filter bubbles and their dangers.
58. See Mohseni and Williams (n.d.).
59. A. I. Goldman and Cox (1996) make a related point in far greater analytic detail.
60. See Urofsky and Finkelman (2008).
61. Geislar and Holman (n.d.) have recently proposed criteria to use in evaluating possible interventions in science policy intended to mitigate industrial influence. They emphasize the importance of policies that are responsive to an asymmetric arms race conception of industrial propaganda that Holman previously introduced and that we describe in more detail below.
62. Whether scientists should take this sort of "inductive risk" into account is broadly debated in philosophy of science. See Douglas (2000, 2009).
63. This is not to say that scientists should not publish work that happens to agree with industrial interests (say). Rather, it is that they should weigh the costs of publishing spurious results against the benefits of a flashy paper and demand high standards for research on sensitive topics.
64. There is another side to this consideration, however, which is that consolidating influence in small teams of scientists publishing high-powered studies could, in principle, make it easier for industrial interests because they could focus their attention on these groups.
65. This is reported on by Nuccitelli (2014).
66. The standard is described here: https://en.wikipedia.org/wiki/Wikipedia:Scientific_standards#Proper_weighting. We are grateful to an anonymous reader for Yale University Press for bringing this standard to our attention.
67. Strathern (1997) found that Joe Camel was broadly recognized by children, who associated him with cigarettes. The company was sued and in 1997 decided to settle out of court and end the Joe Camel marketing campaign. *New York Times* coverage of the affair includes Broder (1997).
68. See, for instance, Soergel (2017) or Farand (2017).
69. McDonald-Gibson 2017.
70. See Roxborough (2017) and Schiffrin (2017).
71. McAuley 2018.
72. Some of the legal issues related to this sort of proposal are discussed by Verstraete, Bambauer, and Bambauer (2017).
73. Kitcher 2011, 113.
74. As Kitcher (2001, 2011) points out, there is a widespread identification of democracy with the idea of "free elections"—to have a democratic society, it is both necessary and sufficient to decide by popular vote crucial questions, such as who will write legislation or conduct foreign policy, or in some cases,

what the law of the land will be. But elections are a mechanism, a procedure for realizing something deeper and more important: namely, a way of aggregating the opinions and preferences of a population, or as a means of exercising the "will of the people." An overemphasis on the popular vote as the characteristic feature of democracy obscures the ways in which mere voting is often not an effective means of performing this aggregation. In place of this simplistic picture of democracy-as-majority-rule, Kitcher proposes a conception of democracy grounded in the concepts of freedom and, especially, equality, in the sense that all members of a free and democratic society are equal participants in that freedom. The members of such a society participate in decisions about matters that affect them—and again, they do so as free and equal members of the society, such that no one's preferences dominate by default. The idea is that each of us has interests, things we care about deeply (and things that make little difference to us), but which have an impact on others. In a democratic society, the decisions we make together must be responsive to the interests and values of each of us. The rest of the package—laws, elections, and so forth—are merely imperfect means of realizing these democratic ideals of freedom, equality, and participation.

75. In many ways, Kitcher's primary goal is to understand how a science guided by this sort of deliberative process can and should be responsive to the needs of a diverse society. His targets are the twin ideas that science can truly be pure or free of human concerns, or, if not, that it is nothing but a construct of the cultural context in which it is done. (These, recall, were pillars of the science wars that we discussed in Chapter 1; Kitcher, like us, seeks to find a middle path.) We are strongly sympathetic with much of what Kitcher says.

Bibliography

ABC News. 2006. "20/20: Health Risks of Mercury in Fish." January 6, 2006. http://abcnews.go.com/2020/story?id=124062&page=1.

Abrahamson, Eric, and Lori Rosenkopf. 1997. "Social Network Effects on the Extent of Innovation Diffusion: A Computer Simulation." *Organization Science* 8 (3): 289–309. https://doi.org/10.1287/orsc.8.3.289.

Acemoglue, Daron, Kostas Bimpikis, and Asuman Ozdaglar. 2014. "Communication Information Dynamics in Endogenous Networks." *Theoretical Economics* 9: 41–97.

Acemoglue, Daron, Asuman Ozdaglar, and Ali ParandehGheibi. 2010. "Spread of Misinformation in Social Networks." *Games and Economic Behavior* 70 (2): 194–227.

Advisory Committee to the Surgeon General of the Public Health Service. 1964. "Smoking and Health." Public Health Service Publication No. 1103. https://profiles.nlm.nih.gov/NN/B/B/M/Q/.

Afshar, Mohammad, and Masoud Asadpour. 2010. "Opinion Formation by Informed Agents." *Journal of Artificial Societies and Social Simulation* 13 (4). https://doi.org/10.18564/jasss.1665.

Aisch, Gregor, Jon Huang, and Cecilia Kang. 2016. "Dissecting the #PizzaGate Conspiracy Theories." *New York Times*, December 10, 2016. https://www.nytimes.com/interactive/2016/12/10/business/media/pizzagate.html.

Allcott, Hunt, and Matthew Gentzkow. 2017. "Social Media and Fake News in the 2016 Election." *Journal of Economic Perspectives* 31 (2): 211–236. https://doi.org/10.1257/jep.31.2.211.

Allen, Beth. 1982. "A Stochastic Interactive Model for the Diffusion of Information." *Journal of Mathematical Sociology* 8 (2): 265–281. https://doi.org/10.1080/0022250X.1982.9989925.

Angere, Staffan. n.d. "Knowledge in a Social Network." Unpublished manuscript.

Angere, Staffan, George Masterton, and Erik J. Olsson. n.d. "The Epistemology of Social Networks." Unpublished manuscript.

Angere, Staffan, and Erik J. Olsson. 2017. "Publish Late, Publish Rarely!: Network Density and Group Competence in Scientific Communication." In *Scientific Collaboration and Collective Knowledge: New Essays*, edited by T. Boyer-Kassem, C. Mayo-Wilson, and M. Weisberg. Cambridge: Oxford University Press.

Ariely, Dan. 2008. *Predictably Irrational: The Hidden Forces That Shape Our Decisions.* New York: HarperCollins.

Arifovic, Jasmina, Curtis Eaton, and Graeme Walker. 2015. "The Coevolution of Beliefs and Networks." *Journal of Economic Behavior and Organization* 120: 46–63.

Arrhenius, Svante. 1896. "XXXI. On the Influence of Carbonic Acid in the Air upon the Temperature of the Ground." *Philosophical Magazine* 41 (251): 237–276. https://doi.org/10.1080/14786449608620846.

Asch, Solomon. 1951. "Effects of Group Pressure upon the Modification and Distortion of Judgments." In *Groups, Leadership and Men: Research in Human Relations*, edited by Harold Guetzkow, 222–236. Oxford: Carnegie Press.

"Assessing Russian Activities and Intentions in Recent US Elections." 2017. Office of the Director of National Intelligence, ICA 2017-01D, January 6, 2017. https://www.dni.gov/files/documents/ICA_2017_01.pdf.

Auwaerter, Paul G., John Aucott, and J. Stephen Dumler. 2004. "Lyme Borreliosis (Lyme Disease): Molecular and Cellular Pathobiology and Prospects for Prevention, Diagnosis and Treatment." *Expert Reviews in Molecular Medicine* 6 (2): 1–22. https://doi.org/10.1017/S1462399404007276.

Axelrod, Alan. 2009. *Selling the Great War: The Making of American Propaganda.* New York: St. Martin's.

Axelrod, Robert. 1997. "The Dissemination of Culture: A Model with Local Convergence and Global Polarization." *Journal of Conflict Resolution* 41 (2): 203–226. https://doi.org/10.1177/0022002797041002001.

Bailey, Christopher J. 2016. *US Climate Change Policy.* New York: Routledge.

Baker, Monya. 2016. "1,500 Scientists Lift the Lid on Reproducibility." *Nature News* 533 (7604): 452. https://doi.org/10.1038/533452a.

Bala, Venkatesh, and Sanjeev Goyal. 1998. "Learning from Neighbours." *Review of Economic Studies* 65 (3): 595–621.

Baldassarri, Delia, and Peter Bearman. 2007. "Dynamics of Political Polarization." *American Sociological Review* 72 (5): 784–811. https://doi.org/10.1177/000312240707200507.

Banerjee, Abhijit. 1992. "A Simple Model of Herd Behavior." *Quarterly Journal of Economics* 107 (3): 797–817.

Barker, Gillian, and Philip Kitcher. 2013. *Philosophy of Science: A New Introduction.* New York: Oxford University Press.

Baron, Robert, Joseph A. Vandello, and Bethany Brunsman. 1996. "The Forgotten Variable in Conformity Research: Impact of Task Importance on Social Influence." *Journal of Personality and Social Psychology* 71 (5): 915–927.

Barrett, Jeffrey, Brian Skyrms, and Aydin Mohseni. 2017. "Self-Assembling Networks." *British Journal for the Philosophy of Science.* https://doi.org/10.1093/bjps/axx039.

Barthel, Michael, Amy Mitchell, and Jesse Holcomb. 2016. "Many Americans Believe Fake News Is Sowing Confusion." Pew Research Center. December 15, 2016. http://www.journalism.org/2016/12/15/many-americans-believe-fake-news-is-sowing-confusion/.

BBC News. 2017. Technology. "Brexit: MPs Quiz Facebook over Brexit 'Fake News.'" October 24, 2017. http://www.bbc.com/news/technology-41736333.

Benjamin, Daniel Jacob, James Berger, Magnus Johannesson, Brian A. Nosek, Eric-Jan Wagenmakers, Richard Berk, Kenneth Bollen, et al. 2017. "Redefine Statistical Significance." PsyArXiv Preprints, July 22, 2017. https://doi.org/10.17605/OSF.IO/MKY9J.

Bentzen, Naja. 2017. "'Fake News' and the EU's Response." At a Glance, April 2017. European Parliament. http://www.europarl.europa.eu/RegData/etudes/ATAG/2017/599384/EPRS_ATA(2017)599384_EN.pdf.

Bernays, Edward L. 1923. *Crystallizing Public Opinion.* New York: Boni and Liveright.

———. 1928. *Propaganda.* Reprint, Brooklyn, NY: Ig Publishing, 2005.

———. 1942. "The Marketing of National Policies: A Study of War Propaganda." *Journal of Marketing* 6 (3): 236–244. https://doi.org/10.2307/1245869.

Bhagat, Smriti, Moira Burke, Carlos Diuk, Ismail Onur Filliz, and Sergey Edunov. 2016. "Three and a Half Degrees of Separation." Facebook Research. February 4, 2016. https://research.fb.com/three-and-a-half-degrees-of-separation.

Bikhchandani, Sushil, David Hirshleifer, and Ivo Welch. 1992. "A Theory of Fads, Fashion, Custom, and Cultural Change as Informational Cascades." *Journal of Political Economy* 100 (5): 992–1026. https://doi.org/10.1086/261849.

Bird, Alexander. 2014. *Thomas Kuhn*. New York: Routledge.

Bloch, Matthew, Joseph Keller, and Haeyoun Park. 2015. "Vaccination Rates for Every Kindergarten in California." *New York Times*, February 6, 2015. https://www.nytimes.com/interactive/2015/02/06/us/california-measles-vaccines-map.html.

Bockenstedt, Linda K., Jialing Mao, Emir Hodzic, Stephen W. Barthold, and Durland Fish. 2002. "Detection of Attenuated, Noninfectious Spirochetes in Borrelia Burgdorferi—Infected Mice After Antibiotic Treatment." *Journal of Infectious Diseases* 186 (10): 1430–1437. https://doi.org/10.1086/345284.

Bolin, Bert. 2008. *A History of the Science and Politics of Climate Change: The Role of the Intergovernmental Panel on Climate Change*. Cambridge: Cambridge University Press.

Bond, Richard, and Peter B. Smith. 1996. "Culture and Conformity: A Meta-Analysis of Studies Using Asch's (1952b, 1956) Line Judgement Task." *Psychological Bulletin* 119 (1): 111–137.

Borg, AnneMarie, Daniel Frey, Dunja Šešelja, and Christian Straßer. 2017. "Examining Network Effects in an Argumentative Agent-Based Model of Scientific Inquiry." In *Logic, Rationality, and Interaction*, 391–406. Lecture Notes in Computer Science. Berlin: Springer. https://doi.org/10.1007/978-3-662-55665-8_27.

Bramson, Aaron, Patrick Grim, Daniel J. Singer, William J. Berger, Graham Sack, Steven Fisher, Carissa Flocken, and Bennett Holman. 2016. "Understanding Polarization: Meanings, Measures, and Model Evaluation." *Philosophy of Science* 84 (1): 115–159. https://doi.org/10.1086/688938.

Brandt, Allan. 2009. *The Cigarette Century: The Rise, Fall, and Deadly Persistence of the Product That Defined America*. Reprint edition. New York: Basic Books.

Bratton, Robert L., John W. Whiteside, Michael J. Hovan, Richard L. Engle, and Frederick D. Edwards. 2008. "Diagnosis and Treatment of Lyme Disease." *Mayo Clinic Proceedings* 83 (5): 566–571. https://doi.org/10.4065/83.5.566.

Bright, Liam Kofi. 2017. "Decision Theoretic Model of the Productivity Gap." *Erkenntnis* 82 (2): 421–442. https://doi.org/10.1007/s10670-016-9826-6.

Broder, John M. 1997. "F.T.C. Charges Joe Camel Ad Illegally Takes Aim at Minors." *New York Times*, May 29, 1997. https://www.nytimes.com/1997/05/29/us/ftc-charges-joe-camel-ad-illegally-takes-aim-at-minors.html.

Bromwich, Jonah Engel, and Alan Blinder. 2017. "What We Know About James Alex Fields, Driver Charged in Charlottesville Killing." *New York Times*, August 13, 2017. https://www.nytimes.com/2017/08/13/us/james-alex-fields-charlottesville-driver-.html.

Bruner, Justin, and Cailin O'Connor. 2017. "Power, Bargaining, and Collaboration." In *Scientific Collaboration and Collective Knowledge: New Essays*, edited by T. Boyer-Kassem, C. Mayo-Wilson, and M. Weisberg. Cambridge: Oxford University Press.

Buechel, Berno, Tim Hellmann, and Stefan Klössner. 2015. "Opinion Dynamics and Wisdom Under Conformity." *Journal of Economic Dynamics and Control* 52: 240–257.

Bump, Philip. 2017. "Here's the Public Evidence That Supports the Idea That Russia Interfered in the 2016 Election." *Washington Post*, July 6, 2017. https://www.washingtonpost.com/news/politics/wp/2017/07/06/heres-the-public-evidence-that-supports-the-idea-that-russia-interfered-in-the-2016-election/.

Burgdorfer, W., A. G. Barbour, S. F. Hayes, J. L. Benach, E. Grunwaldt, and J. P. Davis. 1982. "Lyme Disease—A Tick-Borne Spirochetosis?" *Science* 216 (4552): 1317–1319. https://doi.org/10.1126/science.7043737.

Burgess, Alexis G., and John P. Burgess. 2011. *Truth*. Princeton, NJ: Princeton University Press.

Burns, David M., Lora Lee, Larry Z. Shen, Elizabeth Gilpin, H. Dennis Tolley, Jerry Vaughn, and Thomas G. Shanks. 1997. "Cigarette Smoking Behavior in the United States." In *Changes in Cigarette-Related Disease Risks and Their Implication for Prevention and Control*. NCI Tobacco Control Monograph 8, 13–42.

Button, Katherine S., John P. A. Ioannidis, Claire Mokrysz, Brian A. Nosek, Jonathan Flint, Emma S. J. Robinson, and Marcus R. Munafò. 2013. "Power Failure: Why Small Sample Size Undermines the Reliability of Neuroscience." *Nature Reviews Neuroscience* 14 (5): 365–376. https://doi.org/10.1038/nrn3475.

Cameron, Kenzie A., Michael E. Roloff, Elisha M. Friesema, Tiffany Brown, Borko D. Jovanovic, Sara Hauber, and David W. Baker. 2013. "Patient Knowledge and Recall of Health Information Following Exposure to 'Facts and Myths' Message Format Variations." *Patient Education and Counseling* 92 (3): 381–387. https://doi.org/10.1016/j.pec.2013.06.017.

Campbell, Donald T. 1979. "Assessing the Impact of Planned Social Change." *Evaluation and Program Planning* 2 (1):67–90. https://doi.org/10.1016/0149-7189(79)90048-X.

Campbell, W. Joseph. 2001. *Yellow Journalism: Puncturing the Myths, Defining the Legacies*. Westport, CT: Greenwood.

Candia, Julián, and Karina Mazzitello. 2008. "Mass Media Influence Spreading in Social Networks with Community Structure." *Journal of Statistical Mechanics: Theory and Experiment* 2008 (7). http://iopscience.iop.org/article/10.1088/1742-5468/2008/07/P07007/meta.

Carter, K. Codell. 2017. *Childbed Fever: A Scientific Biography of Ignaz Semmelweis.* New York: Routledge.

CAST (Cardiac Arrhythmia Suppression Trial) Investigators. 1989. "Preliminary Report: Effect of Encainide and Flecainide on Mortality in a Randomized Trial of Arrhythmia Suppression After Myocardial Infarction." *New England Journal of Medicine* 321 (6): 406–412.

CDC (Centers for Disease Control and Prevention). 2015 (last reviewed). "Lyme Disease." Last updated January 19, 2018. https://www.cdc.gov/lyme/index.html.

Chamberlin, T. C. 1899. "An Attempt to Frame a Working Hypothesis of the Cause of Glacial Periods on an Atmospheric Basis." *Journal of Geology* 7 (6): 545–584. https://doi.org/10.1086/608449.

Chen, C., K. Wu, V. Srinivasan, and X. Zhang. 2013. "Battling the Internet Water Army: Detection of Hidden Paid Posters." In *Proceedings of the 2013 IEEE/ACM International Conference on Advances in Social Networks Analysis and Mining (ASONAM 2013)*, 116–120. https://doi.org/10.1145/2492517.2492637.

Chen, Xi, Xi Xiong, Minghong Zhang, and Wei Li. 2016. "Public Authority Control Strategy for Opinion Evolution in Social Networks." *Chaos: An Interdisciplinary Journal of Nonlinear Science* 26. https://doi.org/10.1063/1.4960121.

Cheney, Kyle, and Sarah Wheaton. 2016. "The Most Revealing Clinton Campaign Emails in WikiLeaks Release." Politico, October 7, 2016. http://politi.co/2dBOoRl.

Chernow, Ron. 2005. *Alexander Hamilton.* New York: Penguin.

Choi, Hanool, Sang-Hoon Kim, and Jeho Lee. 2010. "Role of Network Structure and Network Effects in Diffusion of Innovations." *Industrial Marketing Management* 39 (1): 170–177. https://doi.org/10.1016/j.indmarman.2008.08.006.

Christakis, Nicholas A., and James H. Fowler. 2007. "The Spread of Obesity in a Large Social Network over 32 Years." *New England Journal of Medicine* 357 (4): 370–379. https://doi.org/10.1056/NEJMsa066082.

———. 2008. "The Collective Dynamics of Smoking in a Large Social Network." *New England Journal of Medicine* 358 (21): 2249–2258. https://doi.org/10.1056/NEJMsa0706154.

———. 2009. *Connected: The Surprising Power of Our Social Networks and How They Shape Our Lives.* New York: Little, Brown.

Cicerone, Ralph J., Richard S. Stolarski, and Stacy Walters. 1974. "Stratospheric Ozone Destruction by Man-Made Chlorofluoromethanes." *Science* 185 (4157): 1165–1167. https://doi.org/10.1126/science.185.4157.1165.

Cillizza, Chris. 2017. "The Tragic Death and Horrible Politicization of Seth Rich,

Explained." CNN Politics, August 2, 2017. http://www.cnn.com/2017/08/02/politics/seth-rich-death-fox-news-trump/index.html.

Clarkson, Thomas W. 1997. "The Toxicology of Mercury." *Critical Reviews in Clinical Laboratory Sciences* 34 (4): 369–403. https://doi.org/10.3109/10408369708998098.

Cloud, David S., Tracy Wilkinson, and Joseph Tanfani. 2017. "FBI Investigates Russian Government Media Organizations Accused of Spreading Propaganda in U.S." *Los Angeles Times*, September 13, 2017. http://www.latimes.com/nation/la-na-russia-propaganda-20170913-story.html.

CNN Library. 2017. "2016 Presidential Campaign Hacking Fast Facts." CNN, October 31, 2017. http://www.cnn.com/2016/12/26/us/2016-presidential-campaign-hacking-fast-facts/index.html.

Cogbill, Charles V., and Gene E. Likens. 1974. "Acid Precipitation in the Northeastern United States." *Water Resources Research* 10 (6): 1133–1137. https://doi.org/10.1029/WR010i006p01133.

Cohen, Jacob. 1992. "A Power Primer." *Psychological Bulletin* 112 (1): 155–159.

Collins, Ben. 2017. "NASA Denies That It's Running a Child Slave Colony on Mars." *Daily Beast*, June 29, 2017. https://www.thedailybeast.com/nasa-denies-that-its-running-a-child-slave-colony-on-mars.

Concha, Joe. 2016. "MSNBC Anchor Apologizes over False Statement About Fox." The Hill, December 9, 2016. http://thehill.com/homenews/media/309763-msnbc-anchor-apologizes-over-false-statement-about-fox.

Condorcet, Jean-Antoine-Nicolas de Caritat, marquis de. 1785. *Essai sur l'application de l'analyse à la probabilité des décisions rendues à la pluralité des voix* . . . Paris. http://gallica.bnf.fr/ark:/12148/bpt6k417181.

Conley, S. T. 1992. "What Do Consumers Think About Irradiated Foods?" *FSIS Food Safety Review* (Fall): 11–15.

Cowan, Ruth Schwartz. 1972. "Francis Galton's Statistical Ideas: The Influence of Eugenics." *Isis* 63 (4): 509–528. https://doi.org/10.1086/351000.

Crossland, Nicholas A., Xavier Alvarez, and Monica E. Embers. 2017. "Late Disseminated Lyme Disease: Associated Pathology and Spirochete Persistence Post-Treatment in Rhesus Macaques." *American Journal of Pathology*. https://doi.org/10.1016/j.ajpath.2017.11.005.

Crutzen, P. J. 1970. "The Influence of Nitrogen Oxides on the Atmospheric Ozone Content." *Quarterly Journal of the Royal Meteorological Society* 96 (408): 320–325. https://doi.org/10.1002/qj.49709640815.

Cummings, K. M., C. P. Morley, and A. Hyland. 2002. "Failed Promises of the

Cigarette Industry and Its Effect on Consumer Misperceptions About the Health Risks of Smoking." *Tobacco Control* 11 (suppl 1): i110–i117. https://doi.org/10.1136/tc.11.suppl_1.i110.

Dabke, Devavrat, and Eva Arroyo. 2016. "Rumors with Personality: A Differential and Agent-Based Model of Information Spread Through Networks." *SIURO* 9: 453–467.

Daley, D. J., and D. G. Kendal. 1965. "Stochastic Rumors." *IMA Journal of Applied Mathematics* 1 (1): 42–55.

Danner, Mark. 2007. "Words in a Time of War." *The Nation*, May 31, 2007. https://www.thenation.com/article/words-time-war/.

Daston, Lorraine, and Peter Galison. 2007. *Objectivity*. New York: Zone Books.

Davidson, Philip W. 2003. "Methylmercury: A Story of Loaves and Fishes." Paper presented at Pollution, Toxic Chemicals, and Mental Retardation: A National Summit. American Association on Mental Retardation, Racine, WI, July 22–24, 2003.

Davidson, Philip W., Gary J. Myers, Christopher Cox, Catherine Axtell, Conrad Shamlaye, Jean Sloane-Reeves, Elsa Cernichiari, et al. 1998. "Effects of Prenatal and Postnatal Methylmercury Exposure from Fish Consumption on Neurodevelopment: Outcomes at 66 Months of Age in the Seychelles Child Development Study." *Journal of the American Medical Association* 280 (8): 701–707. https://doi.org/10.1001/jama.280.8.701.

Davidson, Philip W., Gary J. Myers, Christopher Cox, Conrad Shamlaye, D. O. Marsh, M. A. Tanner, Cheston M. Berlin, J. Sloane-Reeves, E. Cernichiari, and O. Choisy. 1995. "Longitudinal Neurodevelopmental Study of Seychellois Children Following in Utero Exposure to Methylmercury from Maternal Fish Ingestion: Outcomes at 19 and 29 Months." *Neurotoxicology* 16 (4): 677–688.

Davidson, Philip W., Gary J. Myers, and Bernard Weiss. 2004. "Mercury Exposure and Child Development Outcomes." *Pediatrics* 113 (Supplement 3): 1023–1029.

Davis, Devra Lee. 2002. *When Smoke Ran like Water*. New York: Basic Books.

Dawkins, Richard. 2006. *The Selfish Gene: 30th Anniversary Edition*. Oxford: Oxford University Press.

Deffuant, Guillaume. 2006. "Comparing Extremism Propagation Patterns in Continuous Opinion Models." *Journal of Artificial Societies and Social Simulation* 9 (3). http://jasss.soc.surrey.ac.uk/9/3/8.html.

Deffuant, Guillaume, Frédéric Amblard, Gérard Weisbuch, and Thierry Faure. 2002. "How Can Extremism Prevail? A Study Based on the Relative Agreement Interaction Model." *Journal of Artificial Societies and Social Simulation* 5 (4). October 31, 2002. http://jasss.soc.surrey.ac.uk/5/4/1.html.

Deffuant, Guillaume, Sylvie Huet, and Frédéric Amblard. 2005. "An Individual-Based Model of Innovation Diffusion Mixing Social Value and Individual Benefit." *American Journal of Sociology* 110 (4): 1041–1069. https://doi.org/10.1086/430220.

DeMarzo, Peter, Dimitri Vayanos, and Jeffrey Zwiebel. 2003. "Persuasion Bias, Social Influence, and Unidimensional Opinions." *Quarterly Journal of Economics* 118 (3): 909–968.

Dennis, Rutledge M. 1995. "Social Darwinism, Scientific Racism, and the Metaphysics of Race." *Journal of Negro Education* 64 (3): 243–252. https://doi.org/10.2307/2967206.

Department of Homeland Security. 2016. "Joint Statement from the Department of Homeland Security and Office of the Director of National Intelligence on Election Security." October 7, 2016. https://www.dhs.gov/news/2016/10/07/joint-statement-department-homeland-security-and-office-director-national.

Dorn, Harold F. 1954. "The Relationship of Cancer of the Lung and the Use of Tobacco." *American Statistician* 8 (5): 7–13. https://doi.org/10.1080/00031305.1954.10482762.

Douglas, Heather. 2000. "Inductive Risk and Values in Science." *Philosophy of Science* 67 (4): 559–579. https://doi.org/10.1086/392855.

———. 2009. *Science, Policy, and the Value-Free Ideal.* Pittsburgh, PA: University of Pittsburgh Press.

Douven, Igor, and Alexander Riegler. 2010. "Extending the Hegselmann-Krause Model I." *Logic Journal of the IGPL* 18: 323–335.

Downes, Stephen M. 1992. "The Importance of Models in Theorizing: A Deflationary Semantic View." *PSA: Proceedings of the Biennial Meeting of the Philosophy of Science Association* 1992 (1): 142–153. https://doi.org/10.1086/psaprocbienmeetp.1992.1.192750.

Duggins, Peter. 2017. "A Psychologically-Motivated Model of Opinion Change with Applications to American Politics." *Journal of Artificial Societies and Social Simulation* 20(1): 13. http://jasss.soc.surrey.ac.uk/20/1/13/13.pdf.

Dunn, Richard S. 1979. *The Age of Religious Wars, 1559–1715.* 2nd edition. New York: W. W. Norton.

Duret, Claude. 1605. *Histoire admirable des Plantes et Herbes esmerveillables et miraculeuses en nature.* Paris. http://gallica.bnf.fr/ark:/12148/bpt6k5606404b.

Earman, John. 1992. *Bayes or Bust?: A Critical Examination of Bayesian Confirmation Theory.* Cambridge, MA: MIT Press.

Easley, David, and Jon Kleinberg. 2010. *Networks, Crowds, and Markets: Reasoning About a Highly Connected World.* Cambridge: Cambridge University Press.

Easterbrook, Gregg. 1992. "Green Cassandras." *New Republic*, July 6, 1992: 23–25.

Economist. 2007. "The Summer of Acid Rain." *Economist*, December 19, 2007. http://www.economist.com/node/10311405.

Egebark, Johan, and Mathias Ekström. 2011. "Like What You Like or Like What Others Like?: Conformity and Peer Effects on Facebook." IFN Working Paper No. 886. http://www.ifn.se/wfiles/wp/wp886.pdf.

Elliott, Kevin C. 2017. *A Tapestry of Values: An Introduction to Values in Science*. New York: Oxford University Press.

Elliott, Kevin C., and Bennett Holman. n.d. "The Promise and Perils of Industry-Funded Science." Unpublished manuscript.

Ellison, Glenn, and Drew Fudenberg. 1995. "Word-of-Mouth Communication and Social Learning." *Quarterly Journal of Economics* 110 (1): 93–125. https://doi.org/10.2307/2118512.

Ember, Sydney, and Michael M. Grynbaum. 2017. "At CNN, Retracted Story Leaves an Elite Reporting Team Bruised." *New York Times*, September 5, 2017. https://www.nytimes.com/2017/09/05/business/media/cnn-retraction-trump-scaramucci.html.

Embers, Monica E., Stephen W. Barthold, Juan T. Borda, Lisa Bowers, Lara Doyle, Emir Hodzic, Mary B. Jacobs, et al. 2012. "Persistence of Borrelia Burgdorferi in Rhesus Macaques Following Antibiotic Treatment of Disseminated Infection." *PLOS ONE* 7 (1): e29914. https://doi.org/10.1371/journal.pone.0029914.

Embers, Monica E., Nicole R. Hasenkampf, Mary B. Jacobs, Amanda C. Tardo, Lara A. Doyle-Meyers, Mario T. Philipp, and Emir Hodzic. 2017. "Variable Manifestations, Diverse Seroreactivity and Post-Treatment Persistence in Non-Human Primates Exposed to Borrelia Burgdorferi by Tick Feeding." *PLOS ONE* 12 (12): e0189071. https://doi.org/10.1371/journal.pone.0189071.

Engber, Daniel. 2018. "LOL Something Matters." Slate, January 3, 2018. https://slate.com/health-and-science/2018/01/weve-been-told-were-living-in-a-post-truth-age-dont-believe-it.html.

Ewing, Philip. 2016. "WikiLeaks Offers Reward in Search for Democratic Party Staffer's Killer." National Public Radio, August 10, 2016. http://www.npr.org/sections/thetwo-way/2016/08/10/489531198/wikileaks-offers-reward-in-search-for-democratic-party-staffers-killer.

Fallon, B. A., and J. A. Nields. 1994. "Lyme Disease: A Neuropsychiatric Illness." *American Journal of Psychiatry* 151 (11): 1571–1583. https://doi.org/10.1176/ajp.151.11.1571.

Fan, Fangqi, and Witold Pedrycz. 2016. "Opinion Evolution Influenced by Informed Agents." *Physica A: Statistical Mechanics and Its Applications* 462: 431–441.

Fandos, Nicholas. 2017. "White House Pushes 'Alternative Facts.' Here Are the Real Ones." *New York Times*, January 22, 2017. https://www.nytimes.com/2017/01/22/us/politics/president-trump-inauguration-crowd-white-house.html.

Farand, Chloe. 2017. "French Social Media Is Being Flooded with Fake News, Ahead of the Election." Independent, April 22, 2017. http://www.independent.co.uk/news/world/europe/french-voters-deluge-fake-news-stories-facebook-twitter-russian-influence-days-before-election-a7696506.html.

Farman, J. C., B. G. Gardiner, and J. D. Shanklin. 1985. "Large Losses of Total Ozone in Antarctica Reveal Seasonal ClO_x/NO_x Interaction." *Nature* 315 (6016): 207–210. https://doi.org/10.1038/315207a0.

FBI (Federal Bureau of Investigation) National Press Office. 2016. "Statement by FBI Director James B. Comey on the Investigation of Secretary Hillary Clinton's Use of a Personal E-Mail System." July 5, 2016. https://www.fbi.gov/news/pressrel/press-releases/statement-by-fbi-director-james-b-comey-on-the-investigation-of-secretary-hillary-clinton2019s-use-of-a-personal-e-mail-system.

Feder, Henry M. Jr., Barbara J. B. Johnson, Susan O'Connell, Eugene D. Shapiro, Allen C. Steere, Gary P. Wormser, and the Ad Hoc International Lyme Disease Group. 2007. "A Critical Appraisal of 'Chronic Lyme Disease.'" *New England Journal of Medicine* 357 (14): 1422–1430. https://doi.org/10.1056/NEJMra072023.

Festinger, Leon. 1962. *A Theory of Cognitive Dissonance*. Stanford, CA: Stanford University Press.

Fiegerman, Seth. 2017. "Twitter Now Losing Users in the U.S." CNN, July 27, 2017. http://money.cnn.com/2017/07/27/technology/business/twitter-earnings/index.html.

Field, Hartry. 1986. "The Deflationary Conception of Truth." In *Fact, Science and Morality*, edited by Graham Macdonald and Crispin Wright, 55–117. Oxford: Blackwell.

Figueira, Álvaro, and Luciana Oliveira. 2017. "The Current State of Fake News: Challenges and Opportunities." *Procedia Computer Science* 121: 817–825. https://doi.org/10.1016/j.procs.2017.11.106.

Fisher, Marc, John Woodrow Cox, and Peter Hermann. 2016. "Pizzagate: From Rumor, to Hashtag, to Gunfire in D.C." *Washington Post*, December 6, 2016. https://www.washingtonpost.com/local/pizzagate-from-rumor-to-hashtag-to-gunfire-in-dc/2016/12/06/4c7def50-bbd4-11e6-94ac-3d324840106c_story.html.

Folkenflik, David. 2017. "No Apology, No Explanation: Fox News and the Seth Rich Story." National Public Radio, September 15, 2017. http://www.npr.org

/2017/09/15/551163406/fox-news-has-yet-to-explain-what-what-wrong-in-seth-rich-story.

Ford, Matthew. 2017. "Trump's Press Secretary Falsely Claims: 'Largest Audience Ever to Witness an Inauguration, Period.'" *The Atlantic*, January 21, 2017. https://www.theatlantic.com/politics/archive/2017/01/inauguration-crowd-size/514058/.

Förster, Manuel, Ana Mauleon, and Vincent Vannetelbosch. 2016. "Trust and Manipulation in Social Networks." *Network Science* 4 (2): 216–243.

Foucault, Michel. 2012. *The Birth of the Clinic.* New York: Routledge.

Franklin, Ben A. 1984. "Legislators Say White House Suppressed Acid Rain Report." *New York Times*, August 18, 1984. http://www.nytimes.com/1984/08/18/us/legislators-sat-white-house-suppressed-acid-rain-report.html.

Franklin, Benjamin. 1785. "Meteorological Imaginations and Conjectures." *Memoirs of the Literary and Philosophical Society of Manchester*, 2: 357—361.

Freeze, R. Allan, and Jay H. Lehr. 2009. *The Fluoride Wars: How a Modest Public Health Measure Became America's Longest Running Political Melodrama.* Hoboken, NJ: John Wiley & Sons.

Fuller, Steve. 1988. *Social Epistemology.* Bloomington: Indiana University Press.

Funk, Cary, and Lee Rainie. 2015. "Public and Scientists' Views on Science and Society." Pew Research Center, January 29, 2015. http://www.pewinternet.org/2015/01/29/public-and-scientists-views-on-science-and-society/.

Galam, Serge. 2010. "Public Debates Driven by Incomplete Scientific Data: The Cases of Evolution Theory, Global Warming and H1N1 Pandemic Influenza." *Physica A: Statistical Mechanics and Its Applications* 389 (17): 3619–3631. https://doi.org/10.1016/j.physa.2010.04.039.

———. 2011. "Collective Beliefs Versus Individual Inflexibility: The Unavoidable Biases of a Public Debate." *Physica A: Statistical Mechanics and Its Applications* 390 (17): 3036–3054. https://doi.org/10.1016/j.physa.2011.03.021.

Galam, Serge, and Serge Moscovici. 1991. "Towards a Theory of Collective Phenomena: Consensus and Attitude Changes in Groups." *European Journal of Social Psychology* 21 (1): 49–74. https://doi.org/10.1002/ejsp.2420210105.

Geislar, Sally, and Bennett Holman. n.d. "Sex Drugs and Money: How to Evaluate Science Policies Intended to Manage Industry Bias." *Philosophy of Science.* Forthcoming. https://www.academia.edu/35666553/Sex_Drugs_and_Money_How_to_Evaluate_Science_Policies_Intended_to_Manage_Industry_Bias.

Gelfert, Axel. 2014. *A Critical Introduction to Testimony.* London: A & C Black.

Gilbert, Margaret. 1992. *On Social Facts.* Princeton, NJ: Princeton University Press.

Glaberson, William. 1988. "Behind Du Pont's Shift on Loss of Ozone Layer." *New*

York Times, March 26, 1988. http://www.nytimes.com/1988/03/26/business/behind-du-pont-s-shift-on-loss-of-ozone-layer.html.

Godfrey-Smith, Peter. 2009. *Theory and Reality: An Introduction to the Philosophy of Science.* Chicago: University of Chicago Press.

Goldman, Adam. 2016. "The Comet Ping Pong Gunman Answers Our Reporter's Questions." *New York Times*, December 7, 2016. https://www.nytimes.com/2016/12/07/us/edgar-welch-comet-pizza-fake-news.html.

Goldman, Alvin I. 1986. *Epistemology and Cognition.* Cambridge, MA: Harvard University Press.

———. 1999. *Knowledge in a Social World.* Oxford: Oxford University Press.

Goldman, Alvin I., and Thomas Blanchard. 2001 (rev. 2015). "Social Epistemology." In *The Stanford Encyclopedia of Philosophy* (Winter 2016 edition), edited by Edward N. Zalta. https://plato.stanford.edu/archives/win2016/entries/epistemology-social/.

Goldman, Alvin I., and James C. Cox. 1996. "Speech, Truth, and the Free Market for Ideas." *Legal Theory* 2 (1): 1–32.

Golub, Benjamin, and Matthew Jackson. 2007. "Naïve Learning in Social Networks: Convergence, Influence, and Wisdom of Crowds." *American Economic Journal: Microeconomics* 2 (1): 112–149.

Gore, Al. 1992. *Earth in the Balance: Ecology and the Human Spirit.* Boston: Houghton Mifflin

Grandjean, Philippe, Pal Weihe, Roberta F. White, and Frodi Debes. 1998. "Cognitive Performance of Children Prenatally Exposed to 'Safe' Levels of Methylmercury." *Environmental Research* 77 (2): 165–172. https://doi.org/10.1006/enrs.1997.3804.

Grann, David. 2001. "Stalking Dr. Steere over Lyme Disease." *New York Times Magazine*, June 17, 2001.

Granovetter, Mark. 1978. "Threshold Models of Collective Behavior." *American Journal of Sociology* 83: 1360–1380.

Granovetter, Mark S. 1973. "The Strength of Weak Ties." *American Journal of Sociology* 78 (6): 1360–1380. https://doi.org/10.1086/225469.

Greenwood, Shannon, Andrew Perrin, and Maeve Duggan. 2016. "Social Media Update 2016." Pew Research Center, November 11, 2016. http://www.pewinternet.org/2016/11/11/social-media-update-2016/.

Grice, Andrew. 2017. "Fake News Handed Brexiteers the Referendum—and Now They Have No Idea What They're Doing." Independent, January 18, 2017. http://www.independent.co.uk/voices/michael-gove-boris-johnson-brexit-eurosceptic-press-theresa-may-a7533806.html.

Grim, Ryan. 2010. "California Pot Initiative Opposed by Beer Industry." HuffPost, September 21, 2010. https://www.huffingtonpost.com/2010/09/21/this-buds-not-for-you-bee_n_732901.html.

Gross, Paul R., and Norman Levitt. 1997. *Higher Superstition: The Academic Left and Its Quarrels with Science.* Baltimore: Johns Hopkins University Press.

Grundy, Isobel. 1999. *Lady Mary Wortley Montagu.* Oxford: Oxford University Press.

Gupta, Aditi, Hemank Lamba, Ponnurangam Kumaraguru, and Anupam Joshi. 2013. "Faking Sandy: Characterizing and Identifying Fake Images on Twitter During Hurricane Sandy." In *Proceedings of the 22nd International Conference on World Wide Web*, 729–736. New York: ACM. https://doi.org/10.1145/2487788.2488033.

Hacking, Ian. 2001. *An Introduction to Probability and Inductive Logic.* Cambridge: Cambridge University Press.

Hajek, Alan. 2008. "Dutch Book Arguments." In *The Handbook of Rational and Social Choice*, edited by Paul Anand, Prasanta K. Pattanaik, and Clemens Puppe, 173–196. Oxford: Oxford University Press.

Halperin, John J. 2008. "Nervous System Lyme Disease." *Infectious Disease Clinics of North America, Tick-borne Diseases, Part I: Lyme Disease*, 22 (2): 261–274. https://doi.org/10.1016/j.idc.2007.12.009.

Haraway, Donna Jeanne. 1989. *Primate Visions: Gender, Race, and Nature in the World of Modern Science.* Hove, UK: Psychology Press.

Harding, Sandra G. 1986. *The Science Question in Feminism.* Ithaca, NY: Cornell University Press.

Heesen, Remco. 2015. "How Much Evidence Should One Collect?" *Philosophical Studies* 172 (9): 2299–2313. https://doi.org/10.1007/s11098-014-0411-z.

———. 2017a. "Academic Superstars: Competent or Lucky?" *Synthese* 194 (11): 4499–4518. https://doi.org/10.1007/s11229-016-1146-5.

———. 2017b. "Communism and the Incentive to Share in Science." *Philosophy of Science* 84 (4): 698–716. https://doi.org/10.1086/693875.

Hegselmann, Rainer, and Ulrich Krause. 2002. "Opinion Dynamics and Bounded Confidence Models, Analysis, and Simulation." *Journal of Artificial Societies and Social Simulation* 5 (3). http://citeseerx.ist.psu.edu/viewdoc/download?doi=10.1.1.454.3597&rep=rep1&type=pdf.

Heidenry, John. 1995. *Theirs Was the Kingdom: Lila and DeWitt Wallace and the Story of the Reader's Digest.* New York: W. W. Norton.

Hersch, Erika I., and Patrick C. Phillips. 2004. "Power and Potential Bias in Field Studies of Natural Selection." *Evolution; International Journal of Organic Evolution* 58 (3): 479–485.

Higgins, Iain Macleod. 2011. *The Book of John Mandeville: With Related Texts.* Indianapolis, IN: Hackett.

Hightower, Jane M. 2011. *Diagnosis: Mercury: Money, Politics, and Poison.* Washington, DC: Island Press.

Holman, Bennett. 2015. "The Fundamental Antagonism: Science and Commerce in Medical Epistemology." PhD diss., University of California, Irvine.

Holman, Bennett, and Justin Bruner. 2017. "Experimentation by Industrial Selection." *Philosophy of Science* 84 (5): 1008–1019. http://www.journals.uchicago.edu/doi/abs/10.1086/694037.

Holman, Bennett, and Justin Bruner. 2015. "The Problem of Intransigently Biased Agents." *Philosophy of Science* 82 (5): 956–968. https://doi.org/10.1086/683344.

Holyst, Janusz, Krzysztof Kacperski, and Frank Schweitzer. 2001. "Social Impact Models of Opinion Dynamics." *Annual Review of Computational Physics* 19: 253–273.

Huang, Chung-Yuan, Pen-Jung Tzou, and Chuen-Tsai Sun. 2011. "Collective Opinion and Attitude Dynamics Dependency on Informational and Normative Social Influences." *Simulation* 87: 875–892.

Hufbauer, Carolyn Revelle. 1992. "Global Warming: What My Father Really Said." *Washington Post*, September 13, 1992. https://www.washingtonpost.com/archive/opinions/1992/09/13/global-warming-what-my-father-really-said/5791977b-74b0-44f8-a40c-c1a5df6f744d/.

Hume, David. 1738. *A Treatise of Human Nature.* Reprint, Mineola, NY: Dover Publications, 2003.

———. 1748. *An Enquiry Concerning Human Understanding: A Critical Edition.* Reprint, Oxford: Clarendon, 2000.

Hunter, Duncan, Meika Foster, Jennifer O. McArthur, Rachel Ojha, Peter Petocz, and Samir Samman. 2011. "Evaluation of the Micronutrient Composition of Plant Foods Produced by Organic and Conventional Agricultural Methods." *Critical Reviews in Food Science and Nutrition* 51 (6): 571–582. https://doi.org/10.1080/10408391003721701.

Ioannidis, Evangelos, and Nikos Varsakelis. 2017. "False Beliefs in Unreliable Knowledge Networks." *Physica A: Statistical Mechanics and Its Applications* 470: 275–295.

Iraq Body Count. 2013. "Total Violent Deaths Including Combatants, 2003–2013. December 31, 2013. https://www.iraqbodycount.org/analysis/reference/announcements/5/.

Isaac, Mike, and Scott Shane. 2017. "Facebook's Russia-Linked Ads Came in Many

Disguises." *New York Times*, October 2, 2017. https://www.nytimes.com/2017/10/02/technology/facebook-russia-ads-.html.

Isaac, Mike, and Daisuke Wakabayashi. 2017. "Russian Influence Reached 126 Million Through Facebook Alone." *New York Times*, October 30, 2017. https://www.nytimes.com/2017/10/30/technology/facebook-google-russia.html.

Jalili, Mahdi. 2013. "Social Power and Opinion Formation in Complex Networks." *Physica A: Statistical Mechanics and Its Applications* 392 (4): 959–966.

———. 2015. "Coevolution of Opinion Formation and Network Dynamics in Complex Networked Systems." In *2015 International Conference on Information Society*, IEEE. London, November 9–11, 2015. https://doi.org/10.1109/i-society.2015.7366863.

Jeffrey, Richard C. 1990. *The Logic of Decision.* Chicago: University of Chicago Press.

Jern, Alan, Kai-min K. Chang, and Charles Kemp. 2014. "Belief Polarization Is Not Always Irrational." *Psychological Review* 121 (2): 206–224. https://doi.org/10.1037/a0035941.

JIFSAN (Joint Institute for Food and Safety and Applied Nutrition). 1999. "Annual Report 1998–1999." https://jifsan.umd.edu/docs/annual_reports/98-99%20Annual%20Report.pdf.

Johnston, Harold. 1971. "Reduction of Stratospheric Ozone by Nitrogen Oxide Catalysts from Supersonic Transport Exhaust." *Science* 173 (3996): 517–522. https://doi.org/10.1126/science.173.3996.517.

Joint FAO/IAEA/WHO Study Group. 1999. *High-Dose Irradiation: Wholesomeness of Food Irradiated with Doses Above 10 KGy.* Geneva: World Health Organization.

Jones, Lanie. 1988. "Ozone Warning: He Sounded Alarm, Paid Heavy Price." *Los Angeles Times*, July 14, 1988. http://articles.latimes.com/1988-07-14/news/mn-8873_1_ozone-layer.

Jönsson, Martin L., Ulrike Hahn, and Erik J. Olsson. 2015. "The Kind of Group You Want to Belong To: Effects of Group Structure on Group Accuracy." *Cognition* 142 (September): 191–204. https://doi.org/10.1016/j.cognition.2015.04.013.

Kahneman, Daniel. 2011. *Thinking, Fast and Slow.* New York: Macmillan.

Kearns, Cristin E., Laura A. Schmidt, and Stanton A. Glantz. 2016. "Sugar Industry and Coronary Heart Disease Research: A Historical Analysis of Internal Industry Documents." *JAMA Internal Medicine* 176 (11): 1680–1685. https://doi.org/10.1001/jamainternmed.2016.5394.

Keller, Allan. 1969. *The Spanish-American War: A Compact History.* Portland, OR: Hawthorne Books.

Kennedy, David M. 1980. *Over Here: The First World War and American Society.* New York: Oxford University Press.

Kenner, Robert, dir. 2014. *Merchants of Doubt.* Participant Media.

Keynes, Milo. 2008. "Balancing Newton's Mind: His Singular Behaviour and His Madness of 1692–93." *Notes and Records* 62 (3): 289–300. https://doi.org/10.1098/rsnr.2007.0025.

Kidd, Mark, and Irvin M. Modlin. 1998. "A Century of Helicobacter Pylori." *Digestion* 59 (1): 1–15.

Kimble, Gregory A. 1978. *How to Use (and Misuse) Statistics.* Englewood Cliffs, NJ: Prentice-Hall.

Kitcher, Philip. 1990. "The Division of Cognitive Labor." *Journal of Philosophy* 87 (1): 5–22. https://doi.org/10.2307/2026796.

———. 2001. *Science, Truth, and Democracy.* New York: Oxford University Press.

———. 2011. *Science in a Democratic Society.* Amherst, NY: Prometheus Books.

Kleinberg, Jon. 2007. "Cascading Behavior in Networks: Algorithmic and Economic Issues." In *Algorithmic Game Theory*, edited by Noam Nisan, Tim Roughgarden, Eva Tardos, and Vijay Vazirani, 613–632. Cambridge: Cambridge University Press.

Kluger, Richard. 1997. *Ashes to Ashes: America's Hundred-Year Cigarette War, the Public Health, and the Unabashed Triumph of Philip Morris.* New York: Vintage.

Knight, Peter, ed. 2003. *Conspiracy Theories in American History: An Encyclopedia.* Santa Barbara, CA: ABC-CLIO.

Koerner, Claudia, and Cora Lewis. 2017. "Here's What We Know About the Man Accused of Killing a Woman at a White Supremacist Rally." BuzzFeed News, August 13, 2017. https://www.buzzfeed.com/claudiakoerner/what-we-know-about-james-alex-fields-charlottesville-crash?utm_term=.qiD39kE7x#.ntzbvpK5W.

Koop, C. Everett. 1996. Foreword to *The Cigarette Papers.* Edited by Stanton A. Glantz, John Slade, Lisa A. Bero, Peter Hanauer, and Deborah E. Barnes, xiii–xv. Berkeley: University of California Press.

Krauss, Clifford. 1998. "The World; Remember Yellow Journalism." *New York Times*, February 15, 1998. https://www.nytimes.com/1998/02/15/weekinreview/the-world-remember-yellow-journalism.html.

Krimsky, Sheldon. 2004. *Science in the Private Interest: Has the Lure of Profits Corrupted Biomedical Research?* Lanham, MD: Rowman & Littlefield.

Kuhn, Deanna, and Joseph Lao. 1996. "Effects of Evidence on Attitudes: Is Polarization the Norm?" *Psychological Science* 7 (2): 115–120. https://doi.org/10.1111/j.1467-9280.1996.tb00340.x.

Kummerfeld, Erich, and Kevin J. S. Zollman. 2015. "Conservatism and the Scientific State of Nature." *British Journal for the Philosophy of Science* 67 (4): 1057–1076.

Kuntz, Marcel. 2012. "Destruction of Public and Governmental Experiments of GMO in Europe." *GM Crops & Food* 3 (4): 258–264. https://doi.org/10.4161/gmcr.21231.

kurtchella. 2016. "The Death of Seth Rich." Reddit. https://www.reddit.com/r/conspiracy/comments/4sejv7/the_death_of_seth_rich/.

La Rocca, C. E., L. A. Braunstein, and F. Vazquez. 2014. "The Influence of Persuasion in Opinion Formation and Polarization." *Europhysics Letters* 106 (4). http://iopscience.iop.org/article/10.1209/0295-5075/106/40004.

Lancaster, Justin. 2006. "The Cosmos Myth: The Real Truth About the Revelle-Gore Story." Internet Archive Wayback Machine. https://web.archive.org/web/20070824182650/http://home.att.net/~espi/Cosmos_myth.html.

Laudan, Larry. 1981. "A Confutation of Convergent Realism." *Philosophy of Science* 48 (1): 19–49. https://doi.org/10.1086/288975.

Lee, Henry. 1887. *The Vegetable Lamb of Tartary: A Curious Fable of the Cotton Plant. To Which Is Added a Sketch of the History of Cotton and the Cotton Trade.* London.

Lemoine, Nathan P., Ava Hoffman, Andrew J. Felton, Lauren Baur, Francis Chaves, Jesse Gray, Qiang Yu, and Melinda D. Smith. 2016. "Underappreciated Problems of Low Replication in Ecological Field Studies." *Ecology* 97 (10): 2554–2561. https://doi.org/10.1002/ecy.1506.

Levinovitz, Alan. 2017. "Trump Supporters Refuse to Believe Their Own Eyes." *Slate*, January 27, 2017. http://www.slate.com/articles/health_and_science/science/2017/01/trump_supporters_think_trump_crowds_are_bigger_even_when_looking_at_photos.html.

Lewontin, Richard C. 1998. "Survival of the Nicest?" *New York Review of Books*, October 22, 1998. http://www.nybooks.com/articles/1998/10/22/survival-of-the-nicest/.

Likens, Gene E. 1999. "The Science of Nature, the Nature of Science: Long-Term Ecological Studies at Hubbard Brook." *Proceedings of the American Philosophical Society* 143 (4): 558–572.

Likens, Gene E., and F. Herbert Bormann. 1974. "Acid Rain: A Serious Regional Environmental Problem." *Science* 184 (4142): 1176–1179. https://doi.org/10.2307/1738257.

Likens, Gene E., F. Herbert Bormann, and Noye M. Johnson. 1972. "Acid Rain." *Environment: Science and Policy for Sustainable Development* 14 (2): 33–40. https://doi.org/10.1080/00139157.1972.9933001.

Lipton, Eric. 2016. "Man Motivated by 'Pizzagate' Conspiracy Theory Arrested in

Washington Gunfire." *New York Times*, December 5, 2016. https://www.nytimes.com/2016/12/05/us/pizzagate-comet-ping-pong-edgar-maddison-welch.html.

Longino, Helen E. 1990. *Science as Social Knowledge: Values and Objectivity in Scientific Inquiry*. Princeton, NJ: Princeton University Press.

———. 2002. *The Fate of Knowledge*. Princeton, NJ: Princeton University Press.

Lovelock, J. E. 1974. "Atmospheric Halocarbons and Stratospheric Ozone." *Nature* 252 (5481): 292–294. https://doi.org/10.1038/252292a0.

Lu, An, Chunhua Sun, and Yezheng Liu. 2017. "The Impact of Community Structure on the Convergence Time of Opinion Dynamics." *Discrete Dynamics in Nature and Society* 2017. https://doi.org/10.1155/2017/9396824.

MacKenzie, Donald. 1999. "Eugenics and the Rise of Mathematical Statistics in Britain." In *Statistics in Society: The Arithmetic of Politics*, edited by Daniel Dorling and Stephen Simpson, 55–61. London: Arnold.

Macy, Michael, James A. Kitts, Andreas Flache, and Steve Benard. 2003. "Polarization in Dynamic Networks: A Hopfield Model of Emergent Structure." In *Dynamic Social Network Modeling and Analysis: Workshop Summary and Papers*, edited by R. Brieger, K. Carley, and P. Pattison, 162–173. Washington, DC: National Academies Press.

Maddy, Penelope. 2007. *Second Philosophy: A Naturalistic Method*. Oxford: Clarendon.

Mandeville, John. 1900. *The Travels of Sir John Mandeville*. New York: Macmillan.

Marques, Adriana. 2008. "Chronic Lyme Disease: A Review." *Infectious Disease Clinics of North America, Tick-borne Diseases, Part I: Lyme Disease* 22 (2): 341–360. https://doi.org/10.1016/j.idc.2007.12.011.

Marques, Adriana, Sam R. Telford, Siu-Ping Turk, Erin Chung, Carla Williams, Kenneth Dardick, Peter J. Krause, et al. 2014. "Xenodiagnosis to Detect Borrelia Burgdorferi Infection: A First-in-Human Study." *Clinical Infectious Diseases* 58 (7): 937–945. https://doi.org/10.1093/cid/cit939.

Marshall, Barry J. 2002. "The Discovery That Helicobacter Pylori, a Spiral Bacterium, Caused Peptic Ulcer Disease." In *Helicobacter Pioneers: First Hand Accounts from the Scientists Who Discovered Helicobacters 1892–1982*, edited by Barry Marshall. 165–202. Hoboken, NJ: Wiley-Blackwell.

Marshall, Barry J., and J. Robin Warren. 1984. "Unidentified Curved Bacilli in the Stomach of Patients with Gastritis and Peptic Ulceration." *Lancet* 323 (8390): 1311–1315. https://doi.org/10.1016/S0140-6736(84)91816-6.

Marx, Karl, and Friedrich Engels. 1906. *Manifesto of the Communist Party*. Chicago: C. H. Kerr.

Mäs, Michael, and Andreas Flache. 2013. "Differentiation Without Distancing. Explaining Bi-Polarization of Opinions Without Negative Influence." *PLOS*

ONE 8 (11). http://journals.plos.org/plosone/article?id=10.1371/journal.pone.0074516#pone.0074516-Myers1.

Matrisciana, Patrick, dir. 1994. *The Clinton Chronicles: An Investigation into the Alleged Criminal Activities of Bill Clinton.* Citizens for Honest Government.

Maxwell, James, and Forrest Briscoe. 1997. "There's Money in the Air: The CFC Ban and DuPont's Regulatory Strategy." *Business Strategy and the Environment* 6: 276–286.

Mayo-Wilson, Conor, Kevin J. S. Zollman, and David Danks. 2011. "The Independence Thesis: When Individual and Social Epistemology Diverge." *Philosophy of Science* 78 (4): 653–677. https://doi.org/10.1086/661777.

McAuley, James. 2018. "France Weighs a Law to Rein in 'Fake News,' Raising Fears for Freedom of Speech." *Washington Post*, January 10, 2018. https://www.washingtonpost.com/world/europe/france-weighs-a-law-to-rein-in-fake-news-raising-fears-for-freedom-of-speech/2018/01/10/78256962-f558-11e7-9af7-a50bc3300042_story.html.

McCurdy, Patrick. 1975. "Fluorocarbons: Still Time for Fair Shake, Not Bum's Rush." *Chemical Week*, July 4, 1975.

McDonald-Gibson, Charlotte. 2017. "The E.U. Agency Fighting Russia's Wildfire of Fake News with a Hosepipe." *Time*, September 11, 2017. http://time.com/4887297/europe-fake-news-east-stratcom-kremlin/.

McElroy, Michael B., and John C. McConnell. 1971. "Nitrous Oxide: A Natural Source of Stratospheric NO." *Journal of the Atmospheric Sciences* 28 (6): 1095–1098.https://doi.org/10.1175/1520-0469(1971)028<1095:NOANSO>2.0.CO;2.

McElroy, Michael B., Steven C. Wofsy, Joyce E. Penner, and John C. McConnell. 1974. "Atmospheric Ozone: Possible Impact of Stratospheric Aviation." *Journal of the Atmospheric Sciences* 31 (1): 287–304. https://doi.org/10.1175/1520-0469(1974)031<0287:AOPIOS>2.0.CO;2.

McGarity, Thomas O., and Wendy Elizabeth Wagner. 2008. *Bending Science: How Special Interests Corrupt Public Health Research.* Cambridge, MA: Harvard University Press.

McGrayne, Sharon Bertsch. 2011. *The Theory That Would Not Die: How Bayes' Rule Cracked the Enigma Code, Hunted Down Russian Submarines, and Emerged Triumphant from Two Centuries of Controversy.* New Haven, CT: Yale University Press.

McMaster, H. R. 2011. *Dereliction of Duty: Johnson, McNamara, the Joint Chiefs of Staff.* New York: HarperCollins.

Merton, Robert K. 1942. "A Note on Science and Democracy." *Journal of Legal and Political Sociology* 1: 115–126.

———. 1968. "The Matthew Effect in Science." *Science* 159 (3810): 56–63.

———. 1973. *The Sociology of Science: Theoretical and Empirical Investigations.* Chicago: University of Chicago Press.

Michaels, David. 2008. *Doubt Is Their Product: How Industry's Assault on Science Threatens Your Health.* Oxford: Oxford University Press.

Michaels, David, and Celeste Monforton. 2005. "Manufacturing Uncertainty: Contested Science and the Protection of the Public's Health and Environment." *American Journal of Public Health* 95 (S1): S39–S48.

Misak, C. J. 2004. *Truth and the End of Inquiry: A Peircean Account of Truth.* Oxford Philosophical Monographs. Oxford: Oxford University Press.

Mohseni, Aydin, and Cole Randall Williams. n.d. "Truth and Conformity on Networks." Unpublished manuscript.

Molina, Mario J., and F. S. Rowland. 1974. "Stratospheric Sink for Chlorofluoromethanes: Chlorine Atom-Catalysed Destruction of Ozone." *Nature* 249 (5460): 810–812. https://doi.org/10.1038/249810a0.

Molteni, Megan. 2017. "Anti-Vaxxers Brought Their War to Minnesota—Then Came Measles." Wired. May 7, 2017. https://www.wired.com/2017/05/anti-vaxxers-brought-war-minnesota-came-measles/.

Montanari, Andrea, and Amin Saberi. 2010. "The Spread of Innovations in Social Networks." *Proceedings of the National Academy of Sciences* 107 (47): 20196–20201. https://doi.org/10.1073/pnas.1004098107.

Moore, Thomas J. 1995. *Deadly Medicine: Why Tens of Thousands of Heart Patients Died in America's Worst Drug Disaster.* New York: Simon & Schuster.

Moray, Robert. 1677. "A Relation Concerning Barnacles, by Sr. Robert Moray, Lately One of His Majesties Council for the Kingdom of Scotland." *Philosophical Transactions* 12 (137): 925–927. https://doi.org/10.1098/rstl.1677.0032.

Morganroth, J., E. L. Michelson, L. N. Horowitz, M. E. Josephson, A. S. Pearlman, and W. B. Dunkman. 1978. "Limitations of Routine Long-Term Electrocardiographic Monitoring to Assess Ventricular Ectopic Frequency." *Circulation* 58 (3): 408–414. https://doi.org/10.1161/01.CIR.58.3.408.

Morton, Joseph. 2016. "WikiLeaks Offers $20,000 Reward for Help Finding Omaha Native Seth Rich's Killer." *Omaha World Herald.* August 11, 2016. http://www.omaha.com/news/crime/wikileaks-offers-reward-for-help-finding-omaha-native-seth-rich/article_cfb287bc-5e98-11e6-ae0c-8b471b8cbfbb.html.

Moynihan, Ray. 2008. "Key Opinion Leaders: Independent Experts or Drug Representatives in Disguise?" *BMJ: British Medical Journal* 336 (7658): 1402–1403. https://doi.org/10.1136/bmj.39575.675787.651.

Murray, Polly. 1996. *The Widening Circle: A Lyme Disease Pioneer Tells Her Story.* New York: St. Martin's.

Myers, G. J., D. O. Marsh, P. W. Davidson, C. Cox, C. F. Shamlaye, M. Tanner, A. Choi, E. Cernichiari, O. Choisy, and T. W. Clarkson. 1995. "Main Neurodevelopmental Study of Seychellois Children Following in Utero Exposure to Methylmercury from a Maternal Fish Diet: Outcome at Six Months." *Neurotoxicology* 16 (4): 653–664.

Newman, M. E. J. 2001. "The Structure of Scientific Collaboration Networks." *Proceedings of the National Academy of Sciences* 98 (2): 404–409. https://doi.org/10.1073/pnas.98.2.404.

Newton, Roger G. 1997. *The Truth of Science: Physical Theories and Reality*. Cambridge, MA: Harvard University Press.

New York Times. 1917. "The Committee of Public Misinformation." *New York Times*, July 7, 1917. http://query.nytimes.com/mem/archive-free/pdf?res=9B0CE2DC133BE03ABC4F53DFB166838C609EDE.

———. 1953a. "Lung Cancer Rise Is Laid to Smoking; Four Medical Reports Agree That Cigarettes Also Cause Circulatory Diseases." *New York Times*, December 9, 1953. https://www.nytimes.com/1953/12/09/archives/lung-cancer-rise-is-laid-to-smoking-four-medical-reports-agree-that.html.

———. 1953b. "Tobacco Stocks Hit by Cancer Reports; Some Drop to Lows for Year After Medical Warnings, but Industry Spokesman Scoffs." *New York Times*, December 10, 1953. https://www.nytimes.com/1953/12/10/archives/tobacco-stocks-hit-by-cancer-reports-some-drop-to-lows-for-year.html.

———. 2004. "From the Editors; The Times and Iraq." *New York Times*, May 26, 2004. https://www.nytimes.com/2004/05/26/world/from-the-editors-the-times-and-iraq.html.

Nicolia, Alessandro, Alberto Manzo, Fabio Veronesi, and Daniele Rosellini. 2014. "An Overview of the Last 10 Years of Genetically Engineered Crop Safety Research." *Critical Reviews in Biotechnology* 34 (1): 77–88. https://doi.org/10.3109/07388551.2013.823595.

Nierenberg, William A. 1992. "Roger Revelle." *Physics Today* 45 (2): 119–120. https://doi.org/10.1063/1.2809551.

"1990 AAAS Annual Meeting." 1989. *Science* 246 (4935): 1313–1327.

"Nope Francis: Reports That His Holiness Has Endorsed Republican Presidential Candidate Donald Trump Originated with a Fake News Web Site." 2016. Snopes.com, July 10, 2016. https://www.snopes.com/pope-francis-donald-trump-endorsement/.

Norr, Roy. 1952. "Cancer by the Carton." *Reader's Digest*, December 1952.

Nowak, Andrzei, Jacek Szamrei, and Bibb Latané. 1990. "From Private Attitude to

Public Opinion: A Dynamic Theory of Social Impact." *Psychological Review* 97 (3): 362–376.

NPR Staff. 2016. "The Reason Your Feed Became an Echo Chamber—And What to Do About It." National Public Radio, July 24, 2016. http://www.npr.org/sections/alltechconsidered/2016/07/24/486941582/the-reason-your-feed-became-an-echo-chamber-and-what-to-do-about-it.

NSF (National Science Foundation). 2014. Science and Engineering Indicators 2014. "Chapter 7. Science and Technology: Public Attitudes and Understanding." https://www.nsf.gov/statistics/seind14/index.cfm/chapter-7/c7h.htm.

Nuccitelli, Dana. 2014. "John Oliver's Viral Video: The Best Climate Debate You'll Ever See." *Guardian*, May 23, 2014. http://www.theguardian.com/environment/climate-consensus-97-per-cent/2014/may/23/john-oliver-best-climate-debate-ever.

Nyhan, Brendan, and Jason Reifler. 2010. "When Corrections Fail: The Persistence of Political Misperceptions." *Political Behavior* 32 (2): 303–330. https://doi.org/10.1007/s11109-010-9112-2.

O'Connor, Cailin, and James Owen Weatherall. 2016. "Black Holes, Black-Scholes, and Prairie Voles: An Essay Review of Simulation and Similarity, by Michael Weisberg." *Philosophy of Science* 83 (4): 613–626. https://doi.org/10.1086/687265.

———. 2017. "Scientific Polarization." Cornell University Library, arXiv.org. December 12, 2017. https://arxiv.org/abs/1712.04561.

Odoric of Pordenone. 2002. *The Travels of Friar Odoric: A 14th-Century Journal of the Blessed.* Cambridge: Eerdmans.

Ohlheiser, Abby. 2016. "Three Days After Removing Human Editors, Facebook Is Already Trending Fake News." *Washington Post*, August 29, 2016. https://www.washingtonpost.com/news/the-intersect/wp/2016/08/29/a-fake-headline-about-megyn-kelly-was-trending-on-facebook/.

Okruhlik, Kathleen. 1994. "Gender and the Biological Sciences." *Canadian Journal of Philosophy* 24 (suppl 1): 21–42. https://doi.org/10.1080/00455091.1994.10717393.

Olby, Robert Cecil. 1974. *The Path to the Double Helix: The Discovery of DNA.* Reprint, Mineola, NY: Dover Publications, 1994.

Olsson, Erik J. 2013. "A Bayesian Simulation Model of Group Deliberation and Polarization." In *Bayesian Argumentation*, edited by Frank Zenker, 113–133. Synthese Library. Dordrecht: Springer. https://doi.org/10.1007/978-94-007-5357-0_6.

Olsson, Erik J., and Aron Vallinder. 2013. "Norms of Assertion and Communication in Social Networks." *Synthese* 190 (13): 2557–2571. https://doi.org/10.1007/s11229-013-0313-1.

Oman, Luke, Alan Robock, Georgiy L. Stenchikov, and Thorvaldur Thordarson. 2006. "High-Latitude Eruptions Cast Shadow over the African Monsoon and the Flow of the Nile." *Geophysical Research Letters* 33 (18). https://doi.org/10.1029/2006GL027665.

Onnela, J.-P., J. Saramäki, J. Hyvönen, G. Szabó, D. Lazer, K. Kaski, J. Kertész, and A.-L. Barabási. 2007. "Structure and Tie Strengths in Mobile Communication Networks." *Proceedings of the National Academy of Sciences* 104 (18): 7332–7336. https://doi.org/10.1073/pnas.0610245104.

Open Science Collaboration. 2015. "Estimating the Reproducibility of Psychological Science." *Science* 349 (6251): aac4716. https://doi.org/10.1126/science.aac4716.

Oreskes, Naomi, and Erik M. Conway. 2010. *Merchants of Doubt: How a Handful of Scientists Obscured the Truth on Issues from Tobacco Smoke to Global Warming*. New York: Bloomsbury.

O'Sullivan, Donie, and Dylan Byers. 2017. "Exclusive: Even Pokémon Go Used by Extensive Russian-Linked Meddling Effort." CNNMoney. October 13, 2017. http://money.cnn.com/2017/10/12/media/dont-shoot-us-russia-pokemon-go/index.html.

Padalia, Divya. 2014. "Conformity Bias: A Fact or an Experimental Artifact?" *Psychological Studies* 59 (3): 223–230. https://doi.org/10.1007/s12646-014-0272-8.

Palmer, E. D. 1954. "Investigation of the Gastric Mucosal Spirochetes of the Human." *Gastroenterology* 27 (2): 218–220.

Pariser, Eli. 2011. *The Filter Bubble: How the New Personalized Web Is Changing What We Read and How We Think*. London: Penguin.

Peirce, Charles Sanders. 1878. "How to Make Our Ideas Clear." *Popular Science Monthly* 12: 286–302.

Pennycook, Gordon, and David G. Rand. 2017. "Who Falls for Fake News? The Roles of Analytic Thinking, Motivated Reasoning, Political Ideology, and Bullshit Receptivity." September 12, 2017. https://ssrn.com/abstract=3023545.

Peterson, Russell Wilbur. 1999. *Rebel with a Conscience*. Newark: University of Delaware Press.

Pinto, Sebastián, Pablo Balenzuela, and Claudio Dorso. 2016. "Setting the Agenda: Different Strategies of a Mass Media in a Model of Cultural Dissemination." *Physica A: Statistical Mechanics and Its Applications* 458: 378–390.

Poe, Edgar Allan. 1844. *The Balloon-Hoax*. Booklassic, 2015.

"Pope Francis Shocks World, Endorses Donald Trump for President, Releases

Statement." 2016. ETF News, September 26, 2016. https://web.archive.org/web/20160929104011/http://endingthefed.com/pope-francis-shocks-world-endorses-donald-trump-for-president-releases-statement.html.

Popper, Karl. 1959. *The Logic of Scientific Discovery*. Reprint, New York: Routledge, 2005.

Proctor, Robert N. 2012. *Golden Holocaust: Origins of the Cigarette Catastrophe and the Case for Abolition*. Berkeley: University of California Press.

Ramsey, Frank P. 1927. "Facts and Propositions." *Proceedings of the Aristotelian Society* 7 (1): 153–170.

———. 1931. "Truth and Probability." In *The Foundations of Mathematics and Other Logical Essays*, edited by R. B. Braithwaite, 156–198. New York: Harcourt, Brace.

Ray, Dixy Lee, and Louis R. Guzzo. 1990. *Trashing the Planet: How Science Can Help Us Deal with Acid Rain, Depletion of the Ozone, and Nuclear Waste (Among Other Things)*. New York: HarperPerennial.

Reed, Peter. 2014. *Acid Rain and the Rise of the Environmental Chemist in Nineteenth-Century Britain: The Life and Work of Robert Angus Smith*. Burlington, VA: Ashgate.

Revelle, Roger, and Hans E. Suess. 1957. "Carbon Dioxide Exchange Between Atmosphere and Ocean and the Question of an Increase of Atmospheric CO2 During the Past Decades." *Tellus* 9 (1): 18–27. https://doi.org/10.1111/j.2153-3490.1957.tb01849.x.

Rice, Deborah C., Rita Schoeny, and Kate Mahaffey. 2003. "Methods and Rationale for Derivation of a Reference Dose for Methylmercury by the U.S. EPA." *Risk Analysis: An Official Publication of the Society for Risk Analysis* 23 (1): 107–115.

Riedel, Stefan. 2005. "Edward Jenner and the History of Smallpox and Vaccination." *Proceedings (Baylor University. Medical Center)* 18 (1): 21–25.

Rifkin, Jeremy. 1980. *Entropy: A New World View*. New York: Viking Adult.

Robb, Amanda. 2017. "Anatomy of a Fake News Scandal." *Rolling Stone*, November 16, 2017. https://www.rollingstone.com/politics/news/pizzagate-anatomy-of-a-fake-news-scandal-w511904.

Rodríguez, Arezky, M. del Castillo-Mussot, and G. J. Vázquez. 2009. "Induced Monoculture in Axelrod Model with Clever Mass Media." *International Journal of Modern Physics C* 20 (8): 1233–1245.

Rogers, Everett M. 2010. *Diffusion of Innovations*. 4th edition. New York: Simon and Schuster.

Romero, Felipe. 2016. "Can the Behavioral Sciences Self-Correct? A Social Epistemic Study." *Studies in History and Philosophy of Science Part A* 60 (Supplement C): 55–69. https://doi.org/10.1016/j.shpsa.2016.10.002.

Rosenberg, Eli. 2017. "Alex Jones Apologizes for Promoting 'Pizzagate' Hoax." *New York Times*, March 25, 2017. https://www.nytimes.com/2017/03/25/business/alex-jones-pizzagate-apology-comet-ping-pong.html.

Rosenstock, Sarita, Justin Bruner, and Cailin O'Connor. 2016. "In Epistemic Networks, Is Less Really More?" *Philosophy of Science* 84 (2): 234–252. https://doi.org/10.1086/690717.

Rosenthal, Robert. 1979. "The File Drawer Problem and Tolerance for Null Results." *Psychological Bulletin* 86 (3): 638–641.

Ross, Andrew. 1996. *Science Wars*. Durham, NC: Duke University Press.

Rowland, F. Sherwood. 1989. "Chlorofluorocarbons and the Depletion of Stratospheric Ozone." *American Scientist* 77 (1): 36–45.

Roxborough, Scott. 2017. "How Europe Is Fighting Back Against Fake News." *Hollywood Reporter*, August 21, 2017. http://www.hollywoodreporter.com/news/how-europe-is-fighting-back-fake-news-1030837.

RT American. 2016. "RT America—July 28, 2016." July 29, 2016. https://www.rt.com/shows/rt-america/353838-rt-america-july282016/.

RT International. 2016. "Wikileaks Offers $20k Reward over Dead DNC Staffer, but Won't Confirm He Leaked Emails." RT International, August 10, 2016. https://www.rt.com/usa/355361-murdered-dnc-staffer-assange/.

Rubin, Mordecai B. 2001. "The History of Ozone. The Schönbein Period, 1839–1868." *Bulletin for the History of Chemistry* 26 (1): 40–56.

———. 2002. "The History of Ozone. II. 1869–1899." *Bulletin for the History of Chemistry* 27 (2): 81–106.

———. 2003. "The History of Ozone. Part III." *Helvetica Chimica Acta* 86 (4): 930–940. https://doi.org/10.1002/hlca.200390111.

———. 2004. "The History of Ozone. IV. The Isolation of Pure Ozone and Determination of Its Physical Properties." *Bulletin for the History of Chemistry* 29 (2): 99–106.

Scargle, Jeffrey D. 2000. "Publication Bias (the 'File-Drawer Problem') in Scientific Inference." *Journal of Scientific Exploration* 14 (1): 91–106.

Schaffner, Brian, and Samantha Luks. 2017. "This Is What Trump Voters Said When Asked to Compare His Inauguration Crowd with Obama's." *Washington Post*, January 25, 2017. https://www.washingtonpost.com/news/monkey-cage/wp/2017/01/25/we-asked-people-which-inauguration-crowd-was-bigger-heres-what-they-said/.

Schiffrin, Anya. "How Europe Fights Fake News." 2017. *Columbia Journalism Review*, October 26, 2017. https://www.cjr.org/watchdog/europe-fights-fake-news-facebook-twitter-google.php.

Schmidt, Anja, Thorvaldur Thordarson, Luke D. Oman, Alan Robock, and Stephen Self. 2012. "Climatic Impact of the Long-Lasting 1783 Laki Eruption: Inapplicability of Mass-Independent Sulfur Isotopic Composition Measurements." *Journal of Geophysical Research: Atmospheres* 117 (D23). https://doi.org/10.1029/2012JD018414.

Schneider, Mike. n.d. "The Non-Epistemic Origins of Research Strongholds." Working manuscript.

Scudder, Harold H. 1949. "Poe's 'Balloon Hoax.'" *American Literature* 21 (2): 179–190. https://doi.org/10.2307/2922023.

Seitz, Frederick. 1996. "A Major Deception on Global Warming." *Wall Street Journal*, June 12, 1996. http://www.wsj.com/articles/SB834512411338954000.

Semmelweis, Ignác Fülöp. 1983. *The Etiology, Concept, and Prophylaxis of Childbed Fever.* Madison: University of Wisconsin Press.

S. Fred Singer v. Justin Lancaster. 1993. Commonwealth of Massachusetts, Middlesex, Superior Court Department, Civil Action No. 93-2219. http://ossfoundation.us/projects/environment/global-warming/myths/revelle-gore-singer-lindzen/cosmos-myth/Lancaster_affidavit.pdf.

Shalby, Colleen. 2017. "How Seth Rich's Death Became an Internet Conspiracy Theory." *Los Angeles Times*, May 24, 2017. http://www.latimes.com/business/hollywood/la-fi-ct-seth-rich-conspiracy-20170523-htmlstory.html.

Shanklin, Jonathan. 2010. "Reflections on the Ozone Hole." *Nature* 465 (7294): 34–35. https://doi.org/10.1038/465034a.

Shao, Chengcheng, Giovanni Luca Ciampaglia, Onur Varol, Alessandro Flammini, and Filippo Menczer. 2017. "The Spread of Misinformation by Social Bots." Cornell University Library, arXiv.org. July 24, 2017. http://arxiv.org/abs/1707.07592.

Shear, Michael D., Adam Goldman, and Emily Cochrane. 2017. "Steve Scalise Among 4 Shot at Baseball Field; Suspect Is Dead." *New York Times*, June 14, 2017. https://www.nytimes.com/2017/06/14/us/steve-scalise-congress-shot-alexandria-virginia.html.

Shrader-Frechette, Kristin. 2014. *Tainted: How Philosophy of Science Can Expose Bad Science.* Oxford: Oxford University Press.

Silver, Nate. 2012. *The Signal and the Noise: Why So Many Predictions Fail, But Some Don't.* London: Penguin.

Silverman, Craig. 2016a. "How the Bizarre Conspiracy Theory Behind 'Pizzagate' Was Spread." BuzzFeed News, November 4, 2016. https://www.buzzfeed.com/craigsilverman/fever-swamp-election.

———. 2016b. "This Analysis Shows How Viral Fake Election News Stories Outperformed Real News on Facebook." BuzzFeed News, November 16, 2016. https://www.buzzfeed.com/craigsilverman/viral-fake-election-news-outperformed-real-news-on-facebook.

Silverman, Craig, and Jeremy Singer-Vine. 2016. "The True Story Behind the Biggest Fake News Hit of the Election." BuzzFeed News, December 16, 2016. https://www.buzzfeed.com/craigsilverman/the-strangest-fake-news-empire.

Singer, Daniel J., Aaron Bramson, Patrick Grim, Bennett Holman, Jiin Jung, Karen Kovaka, Anika Ranginani, and William J. Berger. n.d. "Rational Political and Social Polarization." Unpublished manuscript.

Singer, S. Fred. 1970. "Global Effects of Environmental Pollution." *Eos: Earth and Space Science News* 51 (5): 476–478. https://doi.org/10.1029/EO051i005p00476.

———. 1984. "Acid Rain: A Billion-Dollar Solution to a Million-Dollar Problem?" *Policy Review* 27: 56–58.

———. 1990. "What to Do About Greenhouse Warming." *Environmental Science & Technology* 24 (8): 1138–1139. https://doi.org/10.1021/es00078a607.

———. 1996. "Swedish Academy's Choice of Honorees Signals That Ozone Politics Played a Role." *Scientist*, March 4, 1996.

———. 2003. "The Revelle-Gore Story: Attempted Political Suppression of Science." In *Politicizing Science: The Alchemy of Policymaking*, edited by Michael Gough, 283–297. Washington, DC: George C. Marshall Institute.

Singer, S. Fred, Roger Revelle, and Chauncey Starr. 1991. "What to Do About Greenhouse Warming: Look Before You Leap." *Cosmos* 1 (April): 28–33.

Singer-Vine, Jeremy, and Craig Silverman. 2016. "Most Americans Who See Fake News Believe It, New Survey Says." BuzzFeed News, December 6, 2016. https://www.buzzfeed.com/craigsilverman/fake-news-survey.

Sismondo, Sergio. 2013. "Key Opinion Leaders and the Corruption of Medical Knowledge: What the Sunshine Act Will and Won't Cast Light On." *Journal of Law, Medicine and Ethics* 41 (3): 635–643. https://doi.org/10.1111/jlme.12073.

Skyrms, Brian. 1984. *Pragmatics and Empiricism*. New Haven, CT: Yale University Press.

———. 1986. *Choice and Chance: An Introduction to Inductive Logic*. Belmont, CA: Wadsworth.

Smaldino, Paul E., and Richard McElreath. 2016. "The Natural Selection of Bad Science." *Open Science* 3 (9): 160384. https://doi.org/10.1098/rsos.160384.

Smith, Andrew F. 1994. *The Tomato in America: Early History, Culture, and Cookery*. Champaign: University of Illinois Press.

Smith, Philip J., Susan Y. Chu, and Lawrence E. Barker. 2004. "Children Who Have

Received No Vaccines: Who Are They and Where Do They Live?" *Pediatrics* 114 (1): 187–195. https://doi.org/10.1542/peds.114.1.187.

Smith-Spangler, Crystal, Margaret L. Brandeau, Grace E. Hunter, J. Clay Bavinger, Maren Pearson, Paul J. Eschbach, Vandana Sundaram, et al. 2012. "Are Organic Foods Safer or Healthier Than Conventional Alternatives?: A Systematic Review." *Annals of Internal Medicine* 157 (5): 348–366. https://doi.org/10.7326/0003-4819-157-5-201209040-00007.

Smythe, Ted Curtis. 2003. *The Gilded Age Press, 1865–1900.* Santa Barbara, CA: Praeger.

Soergel, Andrew. 2017. "European Union Taking on 'Almost Overwhelming' Fake News Reports." *US News & World Report*, November 13, 2017. https://www.usnews.com/news/best-countries/articles/2017-11-13/european-union-taking-on-almost-overwhelming-fake-news-reports.

Sokal, Alan D. 1996. "Transgressing the Boundaries: Toward a Transformative Hermeneutics of Quantum Gravity." *Social Text* 46/47: 217–252. https://doi.org/10.2307/466856.

Sokal, Alan, and Jean Bricmont. 1999. *Fashionable Nonsense: Postmodern Intellectuals' Abuse of Science.* New York: Macmillan.

Solon, Olivia, and Sabrina Siddiqui. 2017. "Russia-Backed Facebook Posts 'Reached 126m Americans' During US Election." *Guardian*, October 31, 2017. http://www.theguardian.com/technology/2017/oct/30/facebook-russia-fake-accounts-126-million.

Specter, Michael. 2013. "The Lyme Wars." *New Yorker*, July 1, 2013. https://www.newyorker.com/magazine/2013/07/01/the-lyme-wars.

Stanford, P. Kyle. 2001. "Refusing the Devil's Bargain: What Kind of Underdetermination Should We Take Seriously?" *Philosophy of Science* 68 (S3): S1–S12. https://doi.org/10.1086/392893.

———. 2010. *Exceeding Our Grasp: Science, History, and the Problem of Unconceived Alternatives.* New York: Oxford University Press.

State of Minnesota and Blue Cross Blue Shield of Minnesota v. Philip Morris, Inc. 1998. Transcript of Proceedings, April 15, 1998. http://www.putnampit.com/tobacco/april15transcript.html.

Steere, Allen C., S. E. Malawista, D. R. Snydman, R. E. Shope, W. A. Andiman, M. R. Ross, and F. M. Steele. 1977. "An Epidemic of Oligoarticular Arthritis in Children and Adults in Three Connecticut Communities." *Arthritis and Rheumatology* 20 (1): 7–17.

Stein, Sam. 2016. "The RNC Is Hosting Its Christmas Party This Year at Donald

Trump's Hotel." HuffPost, December 8, 2016. https://www.huffingtonpost.com/entry/rnc-donald-trump-party_us_5848cf6ee4b0f9723d003c70.

Steingrímsson, Jón. 1998. *Fires of the Earth: The Laki Eruption, 1783–1784*. Reykjavik: Nordic Volcanological Institute.

Stevenson, D. S., C. E. Johnson, E. J. Highwood, V. Gauci, W. J. Collins, and R. G. Derwent. 2003. "Atmospheric Impact of the 1783–1784 Laki Eruption: Part I Chemistry Modelling." *Atmospheric Chemistry and Physics* 3 (3): 487–507.

Stokes, Leah, and Amanda Giang. 2017. "Existing Domestic Mercury Regulations." Mercury Science and Policy at MIT. http://mercurypolicy.scripts.mit.edu/blog/?p=106.

Stolarski, R. S., and R. J. Cicerone. 1974. "Stratospheric Chlorine: A Possible Sink for Ozone." *Canadian Journal of Chemistry* 52 (8): 1610–1615. https://doi.org/10.1139/v74-233.

Stoljar, Daniel, and Nic Damnjanovic. 1997 (rev. 2010). "The Deflationary Theory of Truth." In *The Stanford Encyclopedia of Philosophy* (Fall 2014 edition), edited by Edward N. Zalta. Metaphysics Research Lab, Stanford University. https://plato.stanford.edu/entries/truth-deflationary/.

Strathern, Marilyn. 1997. "'Improving Ratings': Audit in the British University System." *European Review* 5 (3): 305–321.

Straubinger, Reinhard K. 2000. "PCR-Based Quantification of Borrelia Burgdorferi Organisms in Canine Tissues over a 500-Day Postinfection Period." *Journal of Clinical Microbiology* 38 (6): 2191–2199.

Straubinger, Reinhard K., Alix F. Straubinger, Brian A. Summers, and Richard H. Jacobson. 2000. "Status of Borrelia Burgdorferi Infection After Antibiotic Treatment and the Effects of Corticosteroids: An Experimental Study." *Journal of Infectious Diseases* 181 (3): 1069–1081. https://doi.org/10.1086/315340.

Strevens, Michael. 2003. "The Role of the Priority Rule in Science." *Journal of Philosophy* 100 (2): 55–79.

———. 2017. "Scientific Sharing: Communism and the Social Contract." In *Scientific Collaboration and Collective Knowledge: New Essays*, edited by T. Boyer-Kassem, C. Mayo-Wilson, and M. Weisberg. Cambridge: Oxford University Press.

Stricker, Raphael B. 2007. "Counterpoint: Long-Term Antibiotic Therapy Improves Persistent Symptoms Associated with Lyme Disease." *Clinical Infectious Diseases* 45 (2): 149–157. https://doi.org/10.1086/518853.

Stuart, Tessa. 2017. "Seth Rich: What You Need to Know About Discredited Fox News Story." *Rolling Stone*, August 1, 2017. http://www.rollingstone.com/culture/what-you-need-to-know-about-fox-news-seth-rich-story-w495383.

Sun, Lena H. 2017a. "Anti-Vaccine Activists Spark a State's Worst Measles Outbreak in Decades." *Washington Post,* May 5, 2017. https://www.washingtonpost.com/national/health-science/anti-vaccine-activists-spark-a-states-worst-measles-outbreak-in-decades/2017/05/04/a1fac952-2f39-11e7-9dec-764dc781686f_story.html.

———. 2017b. "Despite Measles Outbreak, Anti-Vaccine Activists in Minnesota Refuse to Back Down." *Washington Post,* August 21, 2017. https://www.washingtonpost.com/national/health-science/despite-measles-outbreak-anti-vaccine-activists-in-minnesota-refuse-to-back-down/2017/08/21/886cca3e-820a-11e7-ab27-1a21a8e006ab_story.html.

Sun, Marjorie. 1984. "Acid Rain Report Allegedly Suppressed." *Science* 225 (4668): 1374. https://doi.org/10.1126/science.225.4668.1374.

Sunstein, Cass R. 2007. "Of Montreal and Kyoto: A Tale of Two Protocols." *Harvard Environmental Law Review* 31: 1–66.

Surowiecki, James. 2005. *The Wisdom of Crowds.* New York: Anchor Books.

Suskind, Ron. 2004. "Faith, Certainty and the Presidency of George W. Bush." *New York Times Magazine,* October 17, 2004. https://www.nytimes.com/2004/10/17/magazine/faith-certainty-and-the-presidency-of-george-w-bush.html.

Szucs, Denes, and John P. A. Ioannidis. 2017. "Empirical Assessment of Published Effect Sizes and Power in the Recent Cognitive Neuroscience and Psychology Literature." *PLOS Biology* 15 (3): e2000797. https://doi.org/10.1371/journal.pbio.2000797.

Taber, Charles S., Damon Cann, and Simona Kucsova. 2009. "The Motivated Processing of Political Arguments." *Political Behavior* 31 (2): 137–155. https://doi.org/10.1007/s11109-008-9075-8.

Tacchini, Eugenio, Gabriele Ballarin, Marco L. Della Vedova, Stefano Moret, and Luca de Alfaro. 2017. "Some Like It Hoax: Automated Fake News Detection in Social Networks." Cornell University Library, arXiv.org. April 25, 2017. http://arxiv.org/abs/1704.07506.

Thordarson, Th., and S. Self. 1993. "The Laki (Skaftár Fires) and Grímsvötn Eruptions in 1783–1785." *Bulletin of Volcanology* 55 (4): 233–263. https://doi.org/10.1007/BF00624353.

Thordarson, Th., S. Self, N. Oskarsson, and T. Hulsebosch. 1996. "Sulfur, Chlorine, and Fluorine Degassing and Atmospheric Loading by the 1783–1784 AD Laki (Skaftár Fires) Eruption in Iceland." *Bulletin of Volcanology* 58 (2): 205–225.

Thordarson, Th. 2003. "Atmospheric and Environmental Effects of the 1783–1784 Laki Eruption: A Review and Reassessment." *Journal of Geophysical Research* 108 (D1). https://doi.org/10.1029/2001JD002042.

Thornton, Brian. 2000. "The Moon Hoax: Debates About Ethics in 1835 New

York Newspapers." *Journal of Mass Media Ethics* 15 (2): 89–100. https://doi.org/10.1207/S15327728JMME1502_3.

Time. 1953. "Medicine: Beyond Any Doubt." 1953. *Time*, November 30, 1953.

Tobacco Institute, Inc. 1958. *Tobacco and Health* 1 (4), September–October 1958. https://www.industrydocumentslibrary.ucsf.edu/tobacco/docs/#id=jfhg0009.

Tran, Mark. 2016. "WikiLeaks to Publish More Hillary Clinton Emails—Julian Assange." *Guardian*, June 12, 2016. http://www.theguardian.com/media/2016/jun/12/wikileaks-to-publish-more-hillary-clinton-emails-julian-assange.

Trigo, Ricardo M., J. M. Vaquero, and R. B. Stothers. 2010. "Witnessing the Impact of the 1783–1784 Laki Eruption in the Southern Hemisphere." *Climatic Change* 99 (3–4): 535–546. https://doi.org/10.1007/s10584-009-9676-1.

Turner, Erick H., Annette M. Matthews, Eftihia Linardatos, Robert A. Tell, and Robert Rosenthal. 2008. "Selective Publication of Antidepressant Trials and Its Influence on Apparent Efficacy." *New England Journal of Medicine* 358 (3): 252–260. https://doi.org/10.1056/NEJMsa065779.

Tversky, Amos, and Daniel Kahneman. 1974. "Judgment Under Uncertainty: Heuristics and Biases." *Science* 185 (4157): 1124–1131. https://doi.org/10.1126/science.185.4157.1124.

"2016 Election Results: President Live Map by State, Real-Time Voting Updates." 2016. Politico, updated December 13, 2016. https://www.politico.com/2016-election/results/map/president.

"2016 Election Results: State Maps, Live Updates." 2016. CNN Politics. http://www.cnn.com/election/results.

UCSF (University of California San Francisco) Library. n.d. "Tobacco Control Archives, UC San Francisco." OAC: Online Archive of California. Accessed February 6, 2018. http://www.oac.cdlib.org/institutions/UC+San+Francisco::Tobacco+Control+Archives.

Urofsky, Melvin I., and Paul Finkelman. 2008. "Abrams v. United States (1919)." In *Documents of American Constitutional and Legal History*, 666–667. New York: Oxford University Press.

US Census Office. *Census Reports: Twelfth Census of the United States, Taken in the Year 1900*. 1902. Washington, DC: US Census Office.

US Department of Health and Human Services. 2012. "Laws/Policies." BeTobaccoFree.gov, August 8, 2012. https://betobaccofree.hhs.gov/laws/index.html.

Vallinder, Aron, and Erik J. Olsson. 2014. "Trust and the Value of Overconfidence: A Bayesian Perspective on Social Network Communication." *Synthese* 191 (9): 1991–2007. https://doi.org/10.1007/s11229-013-0375-0.

Vann, Michael G. 2003. "Of Rats, Rice, and Race: The Great Hanoi Rat Massacre, an Episode in French Colonial History." *French Colonial History* 4 (1): 191–203. https://doi.org/10.1353/fch.2003.0027.

Vargo, Chris J., Lei Guo, and Michelle A. Amazeen. 2017. "The Agenda-Setting Power of Fake News: A Big Data Analysis of the Online Media Landscape from 2014 to 2016." *New Media and Society*, June 15, 2017. https://doi.org/10.1177/1461444817712086.

Verstraete, Mark, Derek Bambauer, and Jane Bambauer. 2017. "Identifying and Countering Fake News." August 1, 2017. Arizona Legal Studies Discussion Paper No. 17-15. https://ssrn.com/abstract=3007971.

Wagner, Claudia, Silvia Mitter, Christian Koerner, and Markus Strohmaier. 2012. "When Social Bots Attack: Modeling Susceptibility of Users in Online Social Networks." *CEUR Workshop Proceedings: Making Sense of Microposts*, 41–48. http://ceur-ws.org/Vol-838/paper_11.pdf

Wakefield, A. J., S. H. Murch, A. Anthony, J. Linnell, D. M. Casson, M. Malik, M. Berelowitz, et al. 1998. "Retracted: Ileal-Lymphoid-Nodular Hyperplasia, Non-Specific Colitis, and Pervasive Developmental Disorder in Children." *Lancet* 351 (9103): 637–641. https://doi.org/10.1016/S0140-6736(97)11096-0.

Wald, Robert M. 1984. *General Relativity*. Chicago: University of Chicago Press.

Walden, Eric, and Glenn Browne. 2009. "Sequential Adoption Theory: A Theory for Understanding Herding Behavior in Early Adoption of Novel Technologies." *Journal of the Association for Information Systems* 10 (1): 31–62.

Wallace, Tim. 2017. "Crowd Scientists Say Women's March in Washington Had 3 Times as Many People as Trump's Inauguration." *New York Times*, January 22, 2017. https://www.nytimes.com/interactive/2017/01/22/us/politics/womens-march-trump-crowd-estimates.html.

Wall Street Journal. 1982. "Study for EPA Says Acid Rain Primarily Due to Coal Burning." *Wall Street Journal*, November 2, 1982. http://global.factiva.com/redir/default.aspx?P=sa&an=j000000020020326deb2013nb&cat=a&ep=ASE.

Wang, Charmont. 1992. *Sense and Nonsense of Statistical Inference: Controversy: Misuse, and Subtlety*. Boca Raton, FL: CRC Press.

Warner, K. E. 1991. "Tobacco Industry Scientific Advisors: Serving Society or Selling Cigarettes?" *American Journal of Public Health* 81 (7): 839–842.

Warren, J. Robin, and Barry Marshall. 1983. "Unidentified Curved Bacilli on Gastric Epithelium in Active Chronic Gastritis." *Lancet* 321 (8336): 1273–1275.

Watson, James D. 2011. *The Double Helix: A Personal Account of the Discovery of the Structure of DNA*. New York: Simon and Schuster.

Watts, Duncan J., and Steven H. Strogatz. 1998. "Collective Dynamics of 'Small-World' Networks." *Nature* 393 (6684): 440–442. http://dx.doi.org/10.1038/30918.

Weatherall, James Owen. 2013. *The Physics of Wall Street: A Brief History of Predicting the Unpredictable.* Boston: Houghton Mifflin Harcourt.

———. 2016. *Void: The Strange Physics of Nothing.* New Haven, CT: Yale University Press.

Weatherall, James Owen, and Cailin O'Connor. 2018. "Conformity in Scientific Networks." Cornell University Library, arXiv.org. March 27, 2018. http://arxiv.org/abs/1803.09905.

Weatherall, James Owen, Cailin O'Connor, and Justin Bruner. 2018. "How to Beat Science and Influence People." Cornell University Library, arXiv.org. January 4, 2018. https://arxiv.org/abs/1801.01239.

Wegner, Dana. 2001. "New Interpretations of How the USS *Maine* Was Lost." In *Theodore Roosevelt, the U.S. Navy, and the Spanish-American War,* edited by E. J. Marolda, 7–17. The Franklin and Eleanor Roosevelt Institute Series on Diplomatic and Economic History. New York: Palgrave Macmillan.

Weiner, Rachel. 2016. "In Home Town of Alleged Pizzagate Gunman, Shock and Disappointment." *Washington Post,* December 7, 2016. https://www.washingtonpost.com/local/public-safety/in-home-town-of-alleged-pizzagate-shooter-shock-and-disappointment/2016/12/07/814d89ca-bc1a-11e6-94ac-3d324840106c_story.html.

Weisberg, Michael. 2012. *Simulation and Similarity: Using Models to Understand the World.* Oxford: Oxford University Press.

Weisbuch, Gérard, Guillaume Deffuant, Amblard Frédéric, and Jean-Pierre Nadal. 2002. "Meet, Discuss, and Segregate!" *Complexity* 7 (3): 55–63.

Wells, Georgia. 2017. "Twitter Overstated Number of Users for Three Years." *Wall Street Journal,* October 26, 2017. https://www.wsj.com/articles/twitter-overstated-number-of-users-for-three-years-1509015657.

Wells, Ida B. 2014. *Southern Horrors: Lynch Law in All Its Phases.* Auckland: Floating Press.

Wilcher, Marshall E. 1986. "The Acid Rain Debate in North America: 'Where You Stand Depends on Where You Sit.'" *Environmentalist* 6 (4): 289–298. https://doi.org/10.1007/BF02238061.

Will, George F. 1992. "Al Gore's Green Guilt." *Washington Post,* September 3, 1992, A23.

Wilson, Andy Abrahams, dir. 2008. *Under Our Skin: The Untold Story of Lyme Disease.* Open Eye Pictures. http://underourskin.com/film/.

Winkle, R. A. 1978. "Antiarrhythmic Drug Effect Mimicked by Spontaneous Variability of Ventricular Ectopy." *Circulation* 57 (6): 1116–1121. https://doi.org/10.1161/01.CIR.57.6.1116.
Winkler, Erhard M. 1976. "Natural Dust and Acid Rain." *Water, Air, and Soil Pollution* 6 (2): 295–302.
———. 2013. *Stone: Properties, Durability in Man's Environment.* Berlin: Springer Science & Business Media.
Wofsy, Steven C., and Michael B. McElroy. 1974. "HO_x, NO_x, and ClO_x: Their Role in Atmospheric Photochemistry." *Canadian Journal of Chemistry* 52 (8): 1582–1591. https://doi.org/10.1139/v74-230.
Wood, Gordon S. 1993. *The Radicalism of the American Revolution.* Reprint edition. New York: Vintage.
Woolf, Christopher. 2016. "Back in the 1890s, Fake News Helped Start a War." Public Radio International, December 8, 2016. https://www.pri.org/stories/2016-12-08/long-and-tawdry-history-yellow-journalism-america.
Wynder, Ernest L., Evarts A. Graham, and Adele B. Croninger. 1953. "Experimental Production of Carcinoma with Cigarette Tar." *Cancer Research* 13 (12): 855–864.
Yeung, Catherine W. M., and Robert S. Wyer. 2005. "Does Loving a Brand Mean Loving Its Products? The Role of Brand-Elicited Affect in Brand Extension Evaluations." *Journal of Marketing Research* 42 (4): 495–506. https://doi.org/10.1509/jmkr.2005.42.4.495.
Young, H. Peyton. 2001. *Individual Strategy and Social Structure: An Evolutionary Theory of Institutions.* Princeton, NJ: Princeton University Press.
———. 2006a. "The Diffusion of Innovations in Social Networks." In *The Economy as an Evolving Complex System, III: Current Perspectives and Future Directions*, edited by Lawrence E. Blume and Steven N. Durlauf, 267–282. New York: Oxford University Press.
———. 2006b. "Innovation Diffusion in Heterogeneous Populations." December 28, 2006. Center on Social and Economic Dynamics Working Paper No. 45. https://ssrn.com/abstract=1024811.
———. 2011. "The Dynamics of Social Innovation." *Proceedings of the National Academy of Sciences* 108 (Supplement 4): 21285–21291. https://doi.org/10.1073/pnas.1100973108.
Zadronsky, Brandy, Kelly Weill, Katie Zavadski, and Emma Kerr. 2017. "Congressional Shooter Loved Bernie, Hated 'Racist' Republicans, and Beat His Daughter." *Daily Beast*, June 14, 2017. https://www.thedailybeast.com/congressional-shooter-loved-bernie-sanders-hated-racist-and-sexist-republicans.

Zhao, Hai-Lu, Xun Zhu, and Yi Sui. 2006. "The Short-Lived Chinese Emperors." *Journal of the American Geriatrics Society* 54 (8): 1295–1296. https://doi.org/10.1111/j.1532-5415.2006.00821.x.

Zhao, Laijun, Qin Wang, Jingjing Cheng, Yucheng Chen, Jiajia Wang, and Wei Huang. 2011. "Rumor Spreading Model with Consideration of Forgetting Mechanism: A Case of Online Blogging Livejournal." *Physica A: Statistical Mechanics and Its Applications* 390 (13): 2619–2625.

Zollman, Kevin J. S. 2007. "The Communication Structure of Epistemic Communities." *Philosophy of Science* 74 (5): 574–587. https://doi.org/10.1086/525605.

———. 2010a. "The Epistemic Benefit of Transient Diversity." *Erkenntnis* 72 (1): 17. https://doi.org/10.1007/s10670-009-9194-6.

———. 2010b. "Social Structure and the Effects of Conformity." *Synthese* 172 (3): 317–340. https://doi.org/10.1007/s11229-008-9393-8.

———. 2013. "Network Epistemology: Communication in Epistemic Communities." *Philosophy Compass* 8 (1): 15–27. https://doi.org/10.1111/j.1747-9991.2012.00534.x.

———. 2015. "Modeling the Social Consequences of Testimonial Norms." *Philosophical Studies* 172 (9): 2371–2383. https://doi.org/10.1007/s11098-014-0416-7.

Zucker, Marty, Gaetan Chevalier, Clint Ober, Paul J. Mills, Deepak Chopra. 2017. "The Healing Benefits of Grounding the Human Body." HuffPost, May 29, 2017. https://www.huffingtonpost.com/entry/the-healing-benefits-of-grounding-the-human-body_us_592c585be4b07d848fdc058a.

Acknowledgments

We wrote most of this book while we were Visiting Fellows at the Research School for Philosophy at Australian National University. We are grateful to the university for its support, to the academic staff at ANU, and especially to Katie Steele and Rachael Brown for helpful conversations related to this material. Part of the book was also drafted while Weatherall was a visitor at the Perimeter Institute for Theoretical Physics; he is grateful to the institute for its support, and to Rob Spekkens, in particular, for helpful conversations. This material is based upon work supported by the National Science Foundation under Grant No. 1535139. O'Connor is grateful for NSF's support and thanks the graduate student collaborators who have contributed to projects associated with her grant: Hannah Rubin, Aydin Mohseni, Sarita Rosenstock, Calvin Cochran, Mike Schneider, and Travis LaCroix. Special thanks to Calvin Cochran for his heroic literature searching for this book.

The work here has benefited from many conversations and helpful correspondence with friends and colleagues. We are particularly grateful to Anna Alexandrova, Jeff Barrett, Agnes Bolinska, Bianca Bosker, Simon Huttegger, David Malament, Aydin Mohseni, Matt Nguyen, Robert Northcott, Sarita Rosenstock, Hannah Rubin, Eric Schliesser, Michael Schneider, Brian Skyrms, Kyle Stanford, and Kevin Zollman. Parts of this material were presented at the Australian National University, UC Merced, UC San Diego, and UC Irvine; we are grateful to the audiences for helpful comments and questions. Vera Weatherall also gave helpful feedback on a presentation related to this material. In general, thanks to our colleagues and students at the Department of Logic and Philosophy Science at UC Irvine for continuing support and inspiration. Special thanks to Nicole Bourbaki for her ongoing support.

There are a number of scholars, academics, and journalists whose work has consistently inspired us and whose ideas we have drawn on regularly. We attempt to acknowledge and cite the specific instances in which we use their ideas and work, but their influence deserves general recognition as well. Thanks to Naomi Oreskes, Erik Conway, Kevin Zollman, Bennett Holman, Philip Kitcher, Jeffrey Barrett, Kyle Stanford, and Craig Silverman. Several friends and colleagues were kind enough to read a penultimate draft of the manuscript and offer invaluable comments that substantially improved the final version. We are grateful to Jeff Barrett, Bianca Bosker, Liam K. Bright, Craig Callender, Nic Fillion, Nathan Fulton, Bennett Holman, Matt Nguyen, Tina Rivers, Chris Smeenk, and two anonymous readers for Yale University Press.

Justin Bruner deserves special thanks: in addition to offering many helpful ideas and suggestions as we drafted this manuscript, he was a coauthor on the research paper "How to Beat Science and Influence People," in which the models concerning propaganda were first introduced and analyzed. His previous work with Bennett Holman was also especially influential in how we thought about industry propagandists.

We thank our agent, Zoë Pagnamenta, for her help with this project, and Allison Lewis for her comments on the original proposal and her assistance throughout the process. Bill Frucht at Yale University Press offered many important insights that helped to shape the manuscript; we are extremely grateful for his contributions and support. We are indebted to Jessie Dolch, who copyedited the manuscript for Yale University Press, for her many corrections and helpful suggestions.

Finally we are grateful to Sylvia and Dennis O'Connor for their support and to James and Maureen Weatherall for countless hours of childcare and for their support; and to Eve and Vera Weatherall for their hugs, kisses, snuggles, and really quite infantile jokes.

Index

Abedin, Huma, 148
Abrams v. United States (1919), 179
acid rain, 35–45, 53, 131, 193n41, 193n45, 194n47
Acid Rain Peer Review Panel, 38–41, 135
advertising: as industrial propaganda, 10, 98; Joe Camel, 212n67; Marlboro Man, 126, 183; tobacco, 95–96, 100, 125–128, 182–183, 212n67
affinity groups, 171–173
Alkali Act of 1864 (Britain), 36
Allcott, Hunt, 4, 187n8
AlterNet, 152
Amazeen, Michelle A., 211n47
American Association for the Advancement of Science, 131
American Medical Association (AMA), 117–118
Angus, Robert, 36
Antarctica, ozone depletion over, 19, 22–24, 45
antiarrhythmics, 122–123
antibiotics, 60, 65–66, 70, 75
antidepressants, 205n37
anti-vaxxers, 143–144, 208n83
Aristotle, 46
Arrhenius, Svante, 129
arrhythmic suppression hypothesis, 122–123
Arroyo, Eva, 202n79
asbestos, 203n12
Asch, Solomon, 80–81, 84, 86, 90, 202n80
Assange, Julian, 164–165, 167
atmospheric science, 19–22, 24, 36, 45, 129–130, 133. *See also* ozone depletion
atomism, 192n27

authority, 44, 126–128, 134, 136
autism, 143–144

backfire effect, 63, 200n56
Bacon, Francis, 34, 191n17
bacterial theory, 59–64
Bala, Venkatesh, 52
Bala-Goyal models, 53–60, 71, 74, 84–86, 101–102, 120
barnacle geese, 187n4
Barrett, Jeffrey, 139, 208n77
BAS (British Antarctic Survey), 19–23, 26, 190n3
Bayes' rule, 30–31, 51–52, 57–58, 71, 102, 105, 110, 192n22
BBC (British Broadcasting Corporation), 10
belief convergence, 59, 62, 86, 112, 156–157, 208n75
beliefs: and conformity bias, 48, 51–53, 55–56, 58, 61–63; degrees of, 30–31, 51; and evidence, 51, 71; minority beliefs, 178–179; persistence of, 15–16; and polarization, 69–77, 79, 82–85, 87–88, 90–92; and propaganda, 100, 102–107, 112–113, 115, 120, 123, 125–126, 137–138, 141–146; social dynamics of, 18, 151, 157, 168, 172–173, 177–179; spread of, 5–9, 11–12; and truth, 25, 29–30, 42–44. *See also* false beliefs; true beliefs
Beran, Christa, 207n65
Berkeley, George, 191n17
Bernays, Edward, 98–100, 126–127, 204n18
best available evidence, 127, 185
bias. *See* confirmation bias; conformity bias
biased production: and industrial selection, 119–120; risks to propagandists of, 110; and scientific practice, 107; selective sharing compared to, 112, 114, 117, 134; and tobacco industry, 104–105
Bikhchandani, Sushil, 82
Bird, Alexander, 192n25
Black Lives Matter, 169
Blacktivist (Facebook group), 170
Bogart, John B., 155
Bolin, Bert, 136
boomerang effect, 200n56
Bormann, F. Herbert, 36–37
Böttcher, G., 59
Bramson, Aaron, 208n75
Breitbart News, 151–152, 166, 183
Brexit, 5–6, 15, 153
Bricmont, Jean, 194n53
Britain. *See* United Kingdom
British Antarctic Survey (BAS), 19–23, 26, 190n3
British Broadcasting Corporation (BBC), 10
Brookings Institution, 151
Bruner, Justin, 77, 101, 119–125, 196n16, 197n28, 204–205n31, 204n28, 207n59
Bureau of Labor Statistics, 161
Burgdorfer, Willy, 65
Bush, George W., 25
Button, Katherine S., 206n39
BuzzFeed News, 4, 155, 165

Calendar, Guy, 129
California Medical Association, 50

Campbell, Donald, 211n54
Canada, acid rain in, 37
cancer, tobacco use linked to, 10–11, 93–96, 100, 105, 108, 110–111, 114
Cann, Damon, 75
carbon dioxide, 128–130
Cardiac Arrhythmia Suppression Trial, 123
Caroline of Ansbach (Princess of Wales), 140–142, 208n82
Catholic Church, 97
Cato Institute, 151
Center for American Progress, 151
Centers for Disease Control and Prevention (CDC), 66
CFCs. *See* chlorofluorocarbons
Chamberlin, T. C., 129
Chang, Kai-min K., 200n58
Charles XI (king), 2
Chemical Week (industry magazine), 26
Chen, Xi, 208n82
childbed fever, 78–79
children: advertising's effect on, 212n67; and Lyme disease, 65, 68; and mercury poisoning, 118–119; and vaccinations, 140, 143–144
chlorine, 21–24
chlorofluorocarbons (CFCs): chemical industry questioning dangers of, 26–31, 35, 41, 43, 127; ozone depletion caused by, 20–24, 45, 190nn5–6
Choi, Hanool, 209n87
Christakis, Nicholas, 101, 204n26
Cicerone, Ralph, 22
cigarettes. *See* smoking; tobacco industry
climate change, 10, 76, 130–131, 134, 137, 139, 178, 185
Clinton, Bill, 163
Clinton, Hillary: fake news about, 3–5, 169; and "Pizzagate," 147–149; and Rich murder, 162–164
The Clinton Chronicles (film), 163
Clinton Foundation, 4–5
cliques, 56, 87–88, 91, 146, 177
CNN, retraction of stories by, 166, 211n45
Cold War propaganda, 97, 175
colonialism, 33–34
Comet Ping Pong (pizzeria), 147, 149, 154–155, 168
Comey, James B., 3, 4–5, 147–148, 166
Committee on Public Information (CPI), 97–98
communal norms, 69, 158, 180, 205n34. *See also* conformity bias
Condorcet, Marquis de, 82
confirmation bias, 75–76
conformity bias, 16–17, 79–91; and beliefs, 48, 51–53, 55–56, 58, 61–63; and consensus, 58, 63; and evidence, 88–89; and judgments, 81–82, 84, 200n68; models of, 13–18, 50–55, 91, 177; and propaganda, 17, 141–144; as rational response to uncertainty, 202n80; and social networks, 12, 48–49, 52, 55–57, 59, 61–62, 171, 177–178, 209n87
Congressional Baseball Game for Charity, 68–69
Congressional Budget Office, 161
Conley, S. T., 202n81

consensus: on acid rain, 39; and conformity bias, 58, 63; and polarization, 69, 71–72, 74–75; and propaganda, 106; scientific, 26, 29, 31, 37, 52, 55, 75–76, 85–86, 95, 103, 157, 160; and social networks, 159–161, 164; and tobacco industry, 10, 96
Conway, Erik, 39, 43, 95, 110, 193n41, 202n1, 203n12; *Merchants of Doubt*, 10
Conway, Kellyanne, 25, 191n10
Copernicus, 28
Cosmos (journal), 131–133
Cowan, Ruth Schwartz, 33
credit in science, 15, 198n30
Crick, Francis, 50–51
Crutzen, Paul, 20

Dabke, Devavrat, 202n79
Daley, D. J., 202n79
Dalton, John, 192n27
Dana, Charles A., 155
Dannemeyer, William E., 163
data collection, 119, 206n40. *See also* evidence
Davidson, Philip, 118
deliberative process, 213n75
democracy: and deliberative process, 213n75; and fake news, 188n16; and free elections, 212–213n74; and free speech, 18, 179, 183; ideals of, 184–186; and propaganda, 99, 176, 184
Democratic National Committee (DNC), 162–165, 167, 169, 211n35
diffusion of ideas, 196n14
disquotationalism, 188n17
distrust, 73–74, 178
DNA model, 19, 50
DNC. *See* Democratic National Committee
Dr. Strangelove (film), 15
Doll, Richard, 111
Dorn, Harold, 94
Douglas, Heather, 180
"doxing" phenomenon, 207n72
Drudge Report, 149
Duggins, Peter, 201n74
DuPont (chemical manufacturer), 26–27, 31, 191n15
Duret, Claude, 187n4

Easterbrook, Gregg, 132, 133, 137
East StratCom, 183
echo chambers, 16–17, 190n31
ecology, 206n39
Einstein, Albert, 28, 48
elections: and fake news, 3–5, 8, 15, 153; free elections as democratic ideal, 184, 212–213n74; Russian interference in, 17, 163–164, 169–172, 183
Electric Power Research Institute (EPRI), 118–119, 122
Elliott, Kevin C., 205n37
empiricism, 34, 191n17
Engels, Friedrich, 97
environmentalism, 42–43
Environmental Protection Agency (EPA), 37, 48, 49
Environmental Science and Technology (journal), 131
epistemic networks, 13–14, 103–104, 197n22
ETF News (endingthefed.com), 3–5, 148, 166, 173, 174, 188n12

eugenics, 33
European Union: East StratCom created by, 183; fake news as political force in, 5–6, 15. *See also specific countries*
evangelization of peoples. *See* propaganda
evidence: and beliefs, 112, 134; best available, 127, 185; and conformity bias, 88–89; misleading, 63; scientific, 11, 41, 52, 94, 96, 157, 181, 184
evolutionary biology, 206n39
expert opinion, 8, 10, 17, 208n82

Facebook: fake news shared on, 3, 4; and minority beliefs, 178; political groups on, 68, 69; reach of, 154, 187n8; responsibility to stop fake news, 173–174; Russian propaganda on, 169–171; social information received from, 16
Fairness Doctrine, 157–159
fake news: and democratic society, 6, 188n16; history of, 152–155, 188n16; impact of, 15; interventions for, 17–18, 173–174, 176, 183–184; and media agenda, 166–168, 211n47; as propaganda, 9–12
Fake News Challenge, 211n53
false beliefs: and conformity bias, 59, 62–64; model for changing, 51–52; persistence of, 6–9, 16, 176; and polarization, 74, 89–91; and propaganda, 112, 138, 180–181; social dynamics of, 11–15, 91, 156–157, 168, 176, 178, 184, 189n25; and truth, 44. *See also* fake news
Farman, Joe, 19
Faroe Islands, and research on mercury levels in fish, 118
Federal Bureau of Investigation (FBI), 3, 163–165
Federal Communications Commission, 157–159
feminist philosophy, 33–34, 199n52
Fields, James Alex, Jr., 69
filter bubbles, 190n30, 212n57
Food and Drug Administration (FDA), 21, 48, 49, 118
Förster, Manuel, 208n82
Foucault, Michel, 33, 192–193n32
Fowler, James, 101, 204n26
Fox News: and fake news, 165–167; retraction of stories by, 165, 166; and social networks, 151
Francis (pope), 3–4, 8, 11, 148
Franklin, Benjamin, 193n38
Franklin, Rosalind, 195n11
free speech, 18, 179, 183
funding of research: by industry, 101, 104–106, 119, 121–123, 181; and propaganda, 108, 115; and selective sharing, 121–122
Funk, Cary, 202n81

Galton, Francis, 34, 35
Gardiner, Brian, 19
Geislar, Sally, 212n61
Gelfert, Axel, 189n21
General Motors (GM), 83
genetically modified organisms (GMOs), 202n81
Gentzkow, Matthew, 4, 187n8

George C. Marshall Institute, 38, 135
Germany: hate speech on social media regulations in, 18, 183; pollution from, 37
global warming. *See* climate change
GMOs (genetically modified organisms), 202n81
Goebbels, Joseph, 97
Goodhart, Charles, 174
Google, 169, 173–174
Google Scholar, 182
Gore, Al, 130–134, 137, 179; *Earth in the Balance*, 130
Goyal, Sanjeev, 52. *See also* Bala-Goyal models
Granovetter, Mark, 197n18
Gray, Freddie, 170
greenhouse gases, 76, 128–131, 133. *See also* climate change
Gregory XV (pope), 97
Gross, Paul, 194–195n54; *Higher Superstition: The Academic Left and Its Quarrels with Science*, 41–42
Grundy, Isobel, 208n81
Guccifer 2.0, 164
Guo, Lei, 211n47

Hajek, Alan, 192n22
Hannity, Sean, 165–166
Hanoi, Vietnam, rat bounty program in, 175
Hansen, James, 207n72
Harding, Sandra: *The Science Question in Feminism*, 33–34
hate speech, 18, 183
heart attacks, 122
Heartland Institute, 134, 151
Heart of Texas (Facebook page), 170
Heesen, Remco, 206n40
Hegselmann, Rainer, 199n54
Herberstein, Sigismund von, 2
Heritage Foundation, 39, 151
Herschel, John, 153
Heuper, Wilhelm, 203n12
Heyer, Heather, 69
Hightower, Jane, 47–49, 52, 54, 57, 61, 161; *Diagnosis Mercury*, 117–118
Hill, John, 95
Hirshleifer, David, 82
Hodgkinson, James Thomas, 68
Hoefnagel, Joris, 187n4
Holman, Bennett, 77, 101, 119–125, 175, 204n28, 204–205n31, 205n37, 207n59, 207n70, 212n61
Holmes, Oliver Wendell, 179
horned hares, 2, 187n4
Hubbard Brook Experimental Forest, 37, 193n41
humanism, 41–42
Hume, David, 27, 28, 29, 34, 191n17
Hussein, Saddam, 160

Iceland, acid rain in, 35–36
Inadvertent Modification of the Stratosphere (IMOS), 21, 26
independence thesis, 190n28
inductive risk, 27, 212n62
industrial propaganda, 10–11, 98–99, 119–120, 212n61. *See also specific industries and campaigns*
industrial selection, 119–124, 134, 181
industry funding of research, 101, 104–106, 119, 121–123, 181
inferences, 7, 27, 50–51, 83
information cascades, 82–83, 200n69, 201n70, 201n74

Infowars, 149, 152, 153, 166, 183
Intergovernmental Panel on Climate Change (IPCC), 134–136, 161
internet memes, 69, 154, 210n19
interventions: for fake news and propaganda, 17–18, 173–174, 176, 183–184; media role in, 168; and networks, 176–177; and polarization, 91; and selective sharing, 116–117; for social media, 177, 183
intransigently biased agents, 207n59
Ioannidis, John P. A., 206n39
Ipsos Public Affairs, 155
Iraq Body Count project, 160, 210n27
irradiated foods, 202n81

jackalope, 2
Jastrow, Robert, 38, 135
Jeffrey, Dick, 71–72, 76, 138
Jern, Alan, 200n58
Joe Camel advertising campaign, 212n67
John Birch Society, 15
Johnson, Lyndon, 25
Jones, Alex, 149, 153
Jones, Charles Ray, 68, 70, 72
Jonstonus, Joannes, 187n4
journalists: and false beliefs, 157–161, 168, 181–182; and propaganda, 17–18, 104–105, 135; and "science wars," 41. *See also* media
Journal of Spirochetal and Tick-Borne Diseases, 67
judgments: and conformity bias, 81–82, 84, 200n68; and polarization, 92; and propaganda, 125, 127–128; social dynamics of, 168

Kaempfer, Engelbert, 2
Keeling, Charles David, 130
Kelly, Megyn, 5, 173, 174
Kemp, Charles, 200n58
Kendal, D. G., 202n79
key opinion leaders, 138, 142, 207n70
Keyworth, George, 38, 40
Kim, Sang-Hoon, 209n87
Kitcher, Philip, 184–186, 194n52, 198n30, 212–213nn74–75; *Science, Truth, and Democracy; Science in a Democratic Society*, 184
Krause, Ulrich, 199n54
Kucsova, Simona, 75
Kuhn, Deanna, 200n59
Kuhn, Thomas, 31–33, 192n25, 192nn27–28; *The Structure of Scientific Revolutions*, 31
Kummerfeld, Erich, 198n32
Kyoto Protocol, 190n8

Laki volcanic eruption (1783), 35–36, 193n36, 193n38
Lancaster, Justin, 132, 207n65
Lao, Joseph, 200n59
Laputa framework, 197n22
Laudan, Larry, 28
Lawrence Livermore National Laboratory, 190n3
Leblanc soda process, 36
Lee, Jeho, 209n87
Letulle, M., 59
Levitt, Norman, 194–195n54; *Higher Superstition: The Academic Left and Its Quarrels with Science*, 41–42
Lewontin, Richard, 42
LGBT United, 169, 171–172
Liggett & Myers, 203n11

Likens, Gene, 36–37, 38
Lippmann, Walter, 204n18
Little, C. C., 134
Locke, John, 191n17
Los Angeles Times on FBI investigations of RT, 189n23
Lown, Bernard, 122
Luks, Samantha, 80
lung cancer, 10, 93, 96, 105, 110–111, 114
Lyme disease, 64–70, 75, 79, 199n45, 199n53
Lyme Disease Foundation, 67

Macron, Emmanuel, 183
Mad Men (TV show), 126
Maine (ship), 152–153, 209n10
Maitland, Charles, 140
Mandeville, John, 1, 2, 11
manufacturing uncertainty, 203n8
marketplace of ideas, 179–180
Marlboro Man, 126, 183
Marshall, Barry, 59, 60, 64
Marshall Institute. *See* George C. Marshall Institute
Marx, Karl, 97
Mason, Monck, 153
Mauleon, Ana, 208n82
maverick effect, 178–179
McCain, John, 178–179
McElreath, Richard, 206n42
McKenzie, Donald, 33
McNamara, Robert, 25
media: and polarization, 80; and propaganda, 10, 94–96; and social networks, 149, 153, 167–168, 174; state-sponsored, 10. *See also* fake news; journalists; social media
memes, 69, 154, 210n19
mercury poisoning, 6, 8, 46–50, 54–55, 79, 117–119
Merton, Robert K., 192n25, 205n34, 206n53
Michaels, David, 203n8
misinformation, 9–11, 18, 77, 144, 168, 182–184. *See also* fake news; propaganda
mistrust, 73–74, 178
modeling: bacterial theory of ulcers, 59–64; Bala-Goyal models, 53–60, 71, 74, 84–86, 101–102, 120; belief changes, 53–58; conformity bias, 13–18, 50–55, 91, 177; heterogeneity of, 195n13; Lyme disease, 64–70; polarization, 68–77, 83–86, 89, 91, 136–137, 172; propaganda effects, 101–112, 115, 120–121, 123–125, 136–138, 145, 204–205n31; social learning, 13–14; social networks, 150–152, 156–159, 171–172, 176, 210n24; weather, 195n13
Mohseni, Aydin, 88–89, 139, 178, 208n77
Molina, Mario, 20–22, 26, 36, 190nn5–6
Montagu, Edward Wortley, 139
Montagu, Mary Wortley, 139–142, 208nn81–82
Montreal Protocol, 24, 190n8
Moray, Robert, 187n4
Morganroth, Joel, 122
motivated reasoning, 75
MSNBC, 151, 166
Murray, Polly, 64–65
Myers, Gary, 119

National Academy of Sciences, 21, 37, 161, 181
National Aeronautics and Space Administration (NASA), 20, 22–23, 26–27
National Fisheries Institute, 119
National Heart, Lung, and Blood Institute, 123
National Institutes of Health, on Lyme disease, 67
National Oceanic and Atmospheric Administration (NOAA), 23
National Science Foundation, 8, 121
National Tuna Foundation, 119
Natural Resources Defense Council (NRDC), 119
Nature on replicability of scientific results, 116
network models: and conformity bias, 12, 48–49, 52, 55–57, 59, 61–62, 171, 177–178, 209n87; epistemic, 13–14, 103–104, 197n22; nodes, 55–57, 73; and polarization, 71–73, 86, 88, 92, 168, 172–173; and propaganda, 120, 124–125, 141, 143, 145; scientific, 12, 52, 61, 70, 102, 123; star, 143, 172. *See also* social networks
neuroscience, 206n39
New England Medical Center, 65
Newman, M. E. J., 197n18
news, 4–5, 152, 155, 158, 167, 184. *See also* fake news; journalists
Newsweek on Rich and DNC server hack, 165
Newton, Isaac, 28, 30, 32, 34, 46, 48, 191n17
New York Journal on explosion of USS *Maine*, 152–154
New York Sun, fake news in, 153
New York Times: on Committee on Public Information, 98; on Iraq Body Count project, 160; on "Pizzagate," 150; as primary news source, 151, 168; on smoking and lung cancer, 94; on social network use by Russians, 170
New York World on explosion of USS *Maine*, 152–154
Nierenberg, William, 38–40, 135, 193n44
Nieuwsuur (Dutch TV show), 164
Nissan, 83
NOAA (National Oceanic and Atmospheric Administration), 23
Nongovernmental International Panel on Climate Change (NIPCC), 134–135
norms, 69, 158, 180, 205n34. *See also* conformity bias
Norway, acid rain in, 37
NRDC (Natural Resources Defense Council), 119

Obama, Barack: and earmark spending, 177; fake news about, 5; inauguration crowds for, 80
obesity, 98, 204n26
O'Connor, Cailin, 196n16, 197n28, 200n54
Odoric (Italian friar), 1
Oliver, John, 182
Olsson, Erik J., 199n54
Onnela, J.-P., 197n18
opinion dynamics, 196n14

opinion leaders, 138, 142, 207n70
Oreskes, Naomi, 39, 43, 95, 110, 193n41, 202n1, 203n12; *Merchants of Doubt*, 10
ozone depletion, 20–24, 26, 28, 34–35, 42, 44–45, 127

Palmer, E. D., 59–60, 64
Palmer Report, 152
Pariser, Eli, 190n30, 212n57
Pearson, Karl, 34, 35
Perot, Ross, 132
pessimistic meta-induction, 28
Pew Research Center, 154, 155
Pilate, Pontius, 24–26
"Pizzagate," 147–150, 167
Podesta, John, 149, 209n3
Poe, Edgar Allan, 153
polarization, 68–77; and beliefs, 69–77, 79, 82–85, 87–92; and conformity bias, 91; and consensus, 69, 71–72, 74–75; defining, 200n57; and interventions, 91; and judgments, 92; and media, 80; models, 68–77, 83–86, 89, 91, 136–137, 172; political, 69, 72, 76; and propaganda, 137; social factors in, 199n54; and social networks, 71–73, 86, 88, 92, 168, 172–173; transient, 74–75; and trust, 17, 71–75, 77, 78, 81; and uncertainty, 72
politics: and democratic society, 176, 179; and false beliefs, 15; and polarization, 17, 92; and propaganda, 10, 97, 104; and scientific research, 31, 34–35, 41, 45; and social networks, 148, 156, 160–161
PolitiFact, 168, 183
pragmatism, 188n17
Problem of Induction, 27, 29
propaganda, 9–11, 93–146; advertising as, 10, 125–128; of Cold War, 97, 175; and conformity bias, 17, 141–144; industrial, 10–11, 98–99, 119–120, 134, 136–137, 139, 212n61; interventions for, 17–18, 173–174, 176, 183–184; intransigently biased agents, 207n59; and key opinion leaders, 207n70; methods, 97–99; modeling effects of, 101–112, 115, 120–121, 123–125, 136–138, 145, 204–205n31; and scientific research, 128; and social networks, 16–17, 141–143, 156–157, 161, 173–176, 178, 180–185
"proper weighting" standard, 182, 212n66
proudcons.com, 5
psychological warfare, 98, 204n18
psychology, 7, 14–15, 75–76, 206n39. *See also* conformity bias
public opinion, 92, 96, 98, 114, 154, 196n14. *See also* conformity bias; key opinion leaders
puerperal fever, 78–79

Qin Shi Huang (emperor), 46

Rahn, Kenneth, 40
Rainie, Lee, 202n81
Ramsey, Frank P., 192n22
Ray, Dixy Lee, 191n15
Reader's Digest on tobacco use and cancer, 93–94
Reagan, Ronald, 38–40, 43

realism, 191n19
Reddit, 149, 162
Reed, Peter, 193n39
replication, 116
reputation, 128, 134–135, 137–139, 142, 172
research funding: by industry, 101, 104–106, 119, 121–123, 181; and propaganda, 108, 115; and selective sharing, 121–122
Revelle, Roger, 128–133, 135, 137, 142, 145, 179
reverse updating, 200n56
Rich, Seth, 162–167, 210n31
Rifkin, Jeremy, 42
Romero, Felipe, 205n32
Rosenstock, Sarita, 197n28
Rosental, Robert, 205n33
Ross, Andrew, 194n52
Rove, Karl, 25, 191n10
Rowland, Sherwood, 20–22, 26, 31, 36, 38, 40–42, 190nn5–6, 191n14
Royal Society of Canada, 37
RT (media organization), 10, 152, 165, 189n23
Ruhle, Stephanie, 166
rumors, diffusion of, 161, 202n79
Russian interference in US and UK politics, 17, 163–164, 169–172, 183

Sacred Congregation for the Propagation of the Faith, 97
Sanders, Bernie, 162–164, 169
San Francisco Medical Society, 50
Santer, Benjamin D., 135–136, 145
Satellite Ozone Analysis Center, 190n3
Scalise, Steve, 68, 69
Scaramucci, Anthony, 166
Scargle, Jeffrey D., 205n33
Schaffner, Brian, 80
Schneider, Mike, 87, 201n75
Schultz, Debbie Wasserman, 162
science: activism in scientific community, 41–43; atmospheric science, 19–22, 24, 36, 45, 129–130, 133; bacterial theory, 59–64; and beliefs, 8, 69–70, 77; and biased production, 107; conformity bias in, 52, 55, 205n34; consensus in, 26, 29, 31, 37, 52, 55, 75–76, 85–86, 95, 103, 157, 160; cultural and political context of, 33–35; evidence in, 11, 41, 52, 94, 96, 157, 181, 184; funding of research by industry, 101, 104–106, 119, 121–123, 181; networks in, 12, 52, 61, 70, 102, 123; and polarization, 75–76, 85–86; and politics, 31, 34–35, 41, 45; reputation's role in, 128, 134–135, 137–139, 142, 172; research in, 11, 41, 52, 96, 101, 104–106, 108, 113–117, 119, 121–123, 145, 180, 181, 184; and social factors in spread of false beliefs, 12–15; and truth, 19–45
Science on acid rain, 36, 40
"science wars," 41–43, 194n53
scientific objectivity, evolution of concept, 192n30
Scripps Institute of Oceanography, 128, 130, 135
Seitz, Frederick, 38, 135, 136, 139, 142
selective sharing, 110–114, 116–117, 121, 134, 205n37
Semmelweis, Ignaz, 77–79, 83, 84, 86, 89, 141, 159–160, 200n60

Seychelles, and research on mercury levels in fish, 118
Shanklin, Jonathan, 19
Shrader-Frechette, Kristin, 195n57
Silver, Nate: *The Signal and the Noise*, 23
Silverman, Craig, 4, 188n9
Singer, D. J., 199n54
Singer, Fred, 39–41, 131–136, 193–194n47, 207n65
Skaggs, L. Craig, 31
Skyrms, Brian, 139, 208n77
Sloan Kettering Memorial Hospital, 94–96, 110–111
Smaldino, Paul E., 206n42
smallpox, 139–142, 208nn81–82
small world networks, 197n18
Smith, A. F., 189n18
smoking, 10–11, 93–96, 100–101, 105, 111, 114, 127, 181–182. *See also* tobacco industry
"Smoking and Health" (TIRC pamphlet), 96
Snopes.com, 168, 183
social dynamics: and conformity bias, 92; of false beliefs, 11–15, 91, 156–157, 168, 176–178, 184, 189n25; influence of, 83, 91, 152, 171; and propaganda, 16, 17, 102, 144, 146; social epistemology, 189n25; social learning models, 13–14
social media: and fake news, 5, 8; interventions for, 177, 183; and network effects, 154, 170, 173; and propaganda, 16, 18. *See also* Facebook; Twitter
social networks, 147–186; and conformity bias, 12, 48–49, 52, 55–57, 59, 61–62, 171, 177–178, 209n87; and consensus, 159–161, 164; and media, 149, 153, 167–168, 174; modeling, 150–152, 156–159, 171–172, 176, 210n24; and polarization, 71–73, 86, 88, 92, 168, 172–173; and politics, 148, 156, 160–161; and propaganda, 16–17, 141–143, 156–157, 161, 173–176, 178, 180–185
Sokal, Alan D., 194n53
Southern Poverty Law Center, 151
Spanish-American War (1898), 152–153, 209n10
Spicer, Sean, 25, 79–80
spurious research results, 108, 113–117, 145, 180
Sputnik News, 10, 152
Stanford, P. Kyle, 28, 191n19
star networks, 143, 172
statistical power, 23, 55, 116
Steere, Allen, 65–68, 70, 72
Steingrímsson, Jón, 35, 193n36
Stockdale, James, 132
Stolarski, Richard, 22, 26–27
Strathern, Marilyn, 174, 212n67
Strevens, Michael, 198n30
Strogatz, Steven H., 197n18
Subject: Politics (website), 148
Suess, Hans, 128–130
sugar industry, 98–99
Sunstein, Cass R., 190n8
supremepatriot.com, 5
Surowiecki, James: *The Wisdom of Crowds*, 200n69

Suskind, Ron, 25
Sweden, acid rain in, 37
Szucs, Denes, 206n39

Taber, Charles, 75
testimonial conundrum, 8, 189n21
Thompson, Linda, 163
Time magazine on tobacco use and cancer, 94
TIRC. *See* Tobacco Industry Research Committee
Tobacco and Health (Tobacco Institute newsletter), 110–111
tobacco industry: advertising, 95–96, 100, 125–128, 182–183, 212n67; misinformation campaigns, 10, 93–96, 100–101, 203n14; propaganda use by, 98, 100–101, 104–105, 110–111, 136; and social networks, 158, 182. *See also* smoking
Tobacco Industry Research Committee (TIRC), 95–96, 100, 103–105, 108, 134
Tobacco Strategy, 95–96, 100–101, 103–104, 114, 116, 123, 134, 156
tomato worm, 189n18
true beliefs: and conformity bias, 59, 62–63; and polarization, 73, 74, 77, 85–88, 92; and propaganda, 103, 107, 112–113, 117, 123, 138, 141; and social networks, 9, 14–15, 156–158, 179; value of, 29
Trump, Donald: CNN stories about retracted, 166; fake news about, 3–5, 8, 148; and inauguration crowd size controversy, 79–80; and Russian interference in 2016 election, 169–170; supporters' behavior, 83–84, 89–90
Trump International Hotel, 166
trust: and polarization, 17, 71–75, 77, 81, 89; and propaganda, 125–126, 134, 136–139; and social networks, 8–9, 151, 168, 170, 172–173, 178
truth, 19–45; metaphysics of, 188n17; and philosophy of science, 27–31; and scientific community, 19–27; and scientific revolutions, 31–35. *See also* fake news
20/20 (TV program), 47, 161
Twitter: "Pizzagate" allegations spread on, 148; reach of, 154; responsibility to stop fake news, 173–174; Russian propaganda on, 169; social information received from, 16
two-armed bandit problem, 196n17

uncertainty: conformity bias as rational response to, 202n80; manufacturing, 203n8; and polarization, 72; and propaganda, 10–11, 94–95, 111, 133, 138; and truth, 26
unconceived alternatives problem, 191n19
United Kingdom: Alkali Act of 1864, 36; fake news as political force in, 5–6, 15–16; and Iraq War, 160; pollution from, 37; Russian political interference in, 17
United Nations, 24, 161–162
United Nations Economic Commission for Europe, 37–38

United States: acid rain legislation considered by, 38–40; air pollution from, 37; fake news in, 5–6, 15–16, 18, 183; IMOS task force formed by, 21; and Iraq War, 160; propaganda methods developed by, 97–98; Russian interference in elections of, 17, 169–172
Unite the Right rally, 68–69
University of Rochester, 118–119, 122

vaccinations, 140, 143–144, 208n83
Vaccine Safety Council of Minnesota, 144
Vannetelbosch, Vincent, 208n82
Vargo, Chris J., 211n47
variolation of smallpox, 139–142, 208nn81–82
Vegetable Lamb of Tartary, 1–3, 5, 8–9, 11, 15, 63, 79, 89, 92
vulgar democracy, 185–186

Wakefield, Andrew, 144
Wall Street Journal as primary news source, 151, 168
Warren, Robin, 59, 60, 64
Washington Post: on "Pizzagate," 150; on political corruption, 3–4; as primary news source, 151, 168; on Rich and DNC server hack, 165, 167
Washington Times as primary news source, 151
Watson, James, 50–51
Watts, Duncan J., 197n18
weaponization of authority, 133–139
Weatherall, James Owen, 200n54
weather modeling, 195n13
Weiner, Anthony, 148
Welch, Edgar Maddison, 147, 150, 154–155
Welch, Ivo, 82
Wheeler, Rod, 165–166
Who Wants to Be a Millionaire? (TV show), 82
WikiLeaks, 149, 164, 167
Wikipedia, 182
Will, George, 132, 133, 137
Williams, Cole, 88–89, 178
Winkle, Robert, 122, 123
women: and childbed fever, 77–79; and feminist philosophy, 33–34, 199n52; and mercury poisoning, 47; tobacco advertising aimed at, 98; and Trump, 5, 80
Women's March (Washington, D.C.), 80
WTOE 5 News, 5

YourNewsWire.com, 148

Zollman, Kevin, 52, 63–64, 189n21, 197n20, 197nn27–28, 198nn30–32, 200n68
Zollman effect, 63–64